GIS AND THE SOCIAL SCIENCES: THEORY AND APPLICATIONS

GIS AND THE SOCIAL SCIENCES: THEORY AND APPLICATIONS

Editor

Halyna Shwetz

Socialy PRESS
AN IMPRINT OF SCITUS ACADEMICS

GIS and the Social Sciences: Theory and Applications

Edited by **Halyna Shwetz**

This Book is Printed on Acid-Free Paper. ♾

ISBN: 978-1-68117-812-7

Library of Congress Control Number: 2017938843

Socialy Press

(An Imprint of Scitus Academics),

616, Corporate Way, Suite 2, 4766,

Valley Cottage, NY 10989

www.scitusacademics.com

Notice

Printed in United States of America.

PREFACE

Today GIS is a major computer application with uses that range from the management of natural resources by government agencies and corporations, to the operations of utility companies, to support for scientific research and education. The use of GIS has now spread very widely among the sciences, and it is now an accepted tool among all of the disciplines that deal with the surface of the Earth and its human population. Moreover the concept of GIS has evolved substantially, and GIS also claims to be an integrating technology, spanning disciplines and blur ring the distinctions between them, both important prerequisites for any broadly useful research infrastructure. The use of GIS has prompted interest in a number of fundamental issues that are collectively identified as geographic information science.

GIS and the Social Sciences: Theory and Applications is dedicated to GIS research and its applications in the fields of Sociology, Criminology, Public Health, Geography, Anthropology, Political Science, and Environmental Studies. GIS (Geographic Information Systems) has grown in popularity as a powerful tool for spatial analysis in the social sciences. Social Science Data and Software (SSDS) continues to build a collaborative network of expertise, support, and resources for GIS and spatial statistical analysis. SSDS has always been a repository of social science numeric data and recent efforts have expanded its collection of spatial data for research and instruction, again with a focus on the social science. This monograph is a valuable resource for students or any social scientist or practitioner interested in applying GIS technology to his or her work.

TABLE OF CONTENTS

1

GIS-BASED MULTI-CRITERIA ANALYSIS FOR LAND USE SUITABILITY ASSESSMENT IN CITY OF REGINA

Jiapei Chen

Faculty of Engineering and Applied Science, University of Regina, S4S 0A2 Regina, SK, Canada

ABSTRACT

Background

Land use suitability assessment is a key factor in any urban and suburban planning and decision-making processes. The assessment is evaluated by a series of criteria involving socio-economic needs. To deal with the conflicting, disproportionate and multiple criteria, land usage will be characterized with respect to the preferences and importance. Meanwhile, many spatial decision problems can be typically analyzed and interpreted visually by applying GIS for mapping and analysis. Accordingly, this study is to introduce the Multi-criteria decision analysis into the land use suitability analysis along with the existing perspective evolving the role of GIS.

Results

For case study, it is the first time to conduct the identification of the current land use situations in City of Regina by using GIS, combined with multi-criteria analysis ideology for existing condition analysis with three criteria referring to social and economic factors. Finally, the study identified five suitability levels, which has revealed the trends, challenges and prospects of land use analysis for urban extension in City of Regina.

Conclusion

Comparing the simulated land use suitable classes to the existing land use pattern, difference and optionality have been presented. The results are providing valuable information for the urban extension for policy and decision makers. To improve the accuracy and the reliability of the real-world case study, criteria selection and weight assignments ranks the first place. The integrated use of data analysis and Multi-criteria Decision Analysis approach, in a GIS context, resulted in a visible assessment of current land use in Regina.

Keywords: Land use suitability, Multi-criteria decision analysis, GIS, Urban extension, Regina

BACKGROUND

Due to the increasing population and economic growth, human activities have continuous impacts on land use. Those impacts might lead to a series complexities toward environment and land resources development (Huang and Xia 2001). Issues related to population and land use competition has emphasized the need for more effective land use planning and policies. In Canada, in particular, the population has been projected to be 47.7 million by 2036 and 63.8 million by 2061, in comparison to 33.7 million in 2009. This pressure of the population will increase the requirement of land use enormously. Land use suitability assessment is to afford a reasonable and sustainable manner for land resource development. Meanwhile, the growing demand for urbanization, land resources used for a variety of purpose will interact and compete with each other. Rational and sustainable decision support based on the land use suitability assessment is an issue of great concern to governments and land users. Thus, an effective way to assess suitability of the current land use and provide policy support to the future land use is desired.

Previously, various methods of spatial analysis for land use are commonly used in the suitability assessment studies. The problem of land use suitability assessment have often been tackled using multi-criteria decision analysis (MCDA) since 1980s (Antoine et al. 1997, Collins et al. 2001, Kiker et al. 2005, Sharifi et al. 2006, Kunwar et al. 2010). Cheng et. al., has reported an integrated MCDA linear programming approach to support selection of an optimal landfill site (Cheng et al. 2003). To create visualized suitability map for users and decision makers, the integration of MCDA and GIS has been widely promoted for solving spatial problems in urban assessment and planning (Phua and Minowa 2005). (Malczewski 2006) conducted a survey of literature of the GIS-based Multi-criteria analysis from 1990 to 2004. There has been a substantial acceleration in the number of the GIS-MCDA articles published in this field. Joerin et al. (2001) put forward an outranking multi-criteria analysis method and output a set of land use suitability maps incorporating complex criteria integrating several stakeholders' points of view. Meanwhile, some case studies were performed using the method of integrating GIS and multi-criteria analysis. Hanyang lake area located in Wuhan city, China, was studied with this comprehensive method to analyze the suitability of future land use according to specified requirements, preferences and predictions in Yong Liu et al.'s research (Liu et al. 2007). Dai et al. (2001) conducted a study of the urban land use planning based on GIS and multi-criteria analysis method, which was applied in Lanzhou city, and its vicinity in north-western China.

However, the research has seldom focused on the Prairie Provinces in Canada, especially for the city with less proportion of population. For the further development and urbanization, the land use suitability assessment will play an important role. Thus, the primary aim of this study is to provide recommendation for integrated and sustainable plan for land use development referring to accessibility, economic, integration and environment by means of weighting the selected criteria. Since the required criteria are heterogeneous and measured on various scales in this method. In addition, the data is limited, this research will only analyze the current situation and future exploitation forecasting based on simplified factors, such as population and income.

The assessment is often assessed by a number of criteria. To deal with the inconsistent, incommensurate and multiple criteria, a comprehensive method is needed to analysis the land use scheme under current situation. Then it can be applied to the land use forecasting and planning. The individual criteria will be characterized with respect to the significance and preferences. Meanwhile, spatial decision problems can also be typically analyzed and demonstrated on the base of visualization. Accordingly, the study is to introduce the Multi-criteria decision analysis into the land use suitability analysis along with the existing perspective evolving the role of Geographical Information Systems (GIS).In this study, the existing land use situation was analyzed and evaluated. Then, based on the analysis the importance of each factor or criterion, the multi-criteria analysis system for land use suitability will be applied. Finally, comparison between the existing land use pattern and forecasting land use possibility of the same area will be conducted. The flowchart is shown in Figure 1. Provided data sets in order to generate and store them in a GIS framework, while, as far as the last data sets are concerned, MCDA was involved in the calculation and acquisition in a GIS environment.

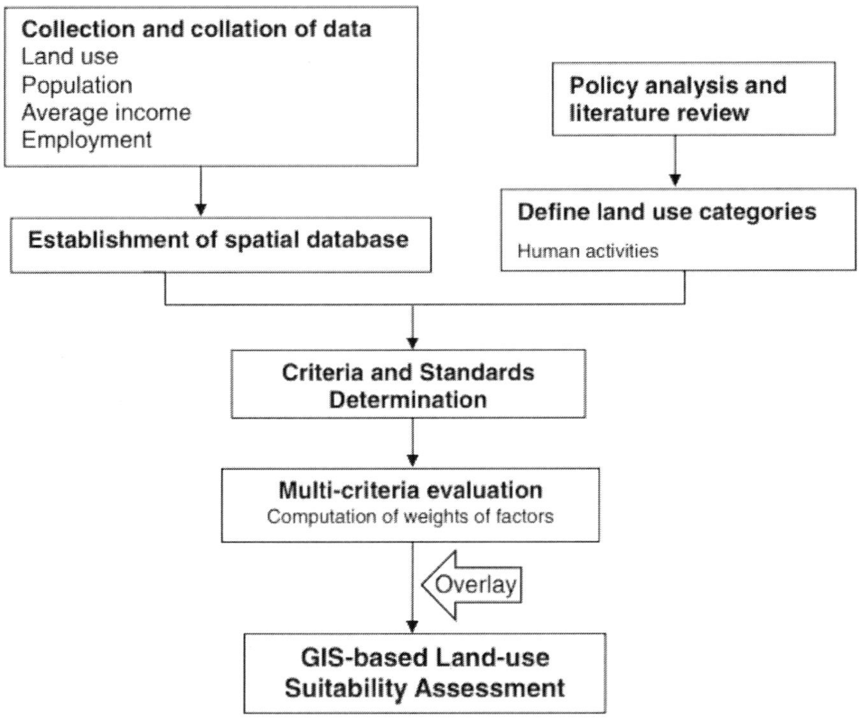

Figure 1: Flowchart of the study.

The study area

The study area is the City of Regina, located in southern Saskatchewan, Canada (as shown in Figure 2). The coordinate of the site is 50°27′17″N, 104°36′24″W. The area of the city is 118.87 km^2. Total population of the study area was 179,246 according to census record in 2006. As the capital city of Saskatchewan, city of Regina ranks the second largest in the province, and is a cultural and commercial metropolis in southern Saskatchewan.The current land use map is presented in Figure 3. There are eight main categories in this study area: Airport, Commerce, Industry, Resident, Institution, Open space, Railway, and Urban holdings. Figure 4 shows the proportion of different land use types. As see form the graph, most of the land in Regina is used as residential area, and then is the open space.

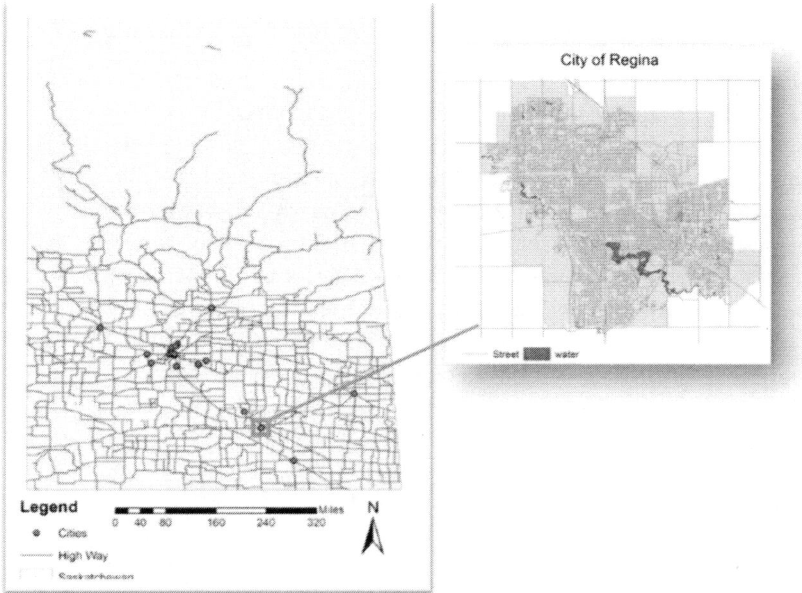

Figure 2: Location of study area and the Saskatchewan Canada.

It is reported that strong economic growth has led to employment and population growth in the Population, Employment and Economic Analysis of city of Regina. Thus population, employment and income are chosen as social and economic effect factors. The data source and information are obtained from TerraServer of TERRA Lab, website of Bureau of Saskatchewan and Canada Statistics. On the basis of "A Guide to the Municipal Planning Process in Saskatchewan", a comprehensive policy framework to guide physical, environmental, economic, social, and cultural development in municipality is provided. As social development is one of the most important factors in land use planning, it is considered separately. The data that will be used in the multi-criteria analysis would be related to population, average income of particular census, and employment.

METHODS

Multi-Criteria Decision Analysis (MCDA), developed in the environmental of Operation Research, aids analysts and decision-makers in situations in which there is a need for identification of priorities according to multiple criteria. This usually happens in situations where conflictive interests coexist (Gomes and Lins 2002). MCDA can incorporate both geographical data and stakeholders' preferences into quantified values for assessment and further decisions (Malczewski 2004). The GIS analysis, if integrated with a procedure of data analysis and structuring, can be usefully developed when data are available but the decision context cannot indicate how these data have to be used to produce information and support decisions. The GIS support the solution of complex spatial problems, providing the decision-maker with a flexible environment in the process of the decision research and in the solution of the problem. The visualization of the context, structure of the problem and its alternative solutions is one for the most powerful components of a decision support system (Gomes and Lins 2002). Thus the integration GIS-MCDA has the objective of the supporting decision-makers, providing them with ways to evaluate several alternatives, based on multiple, conflictive criteria.

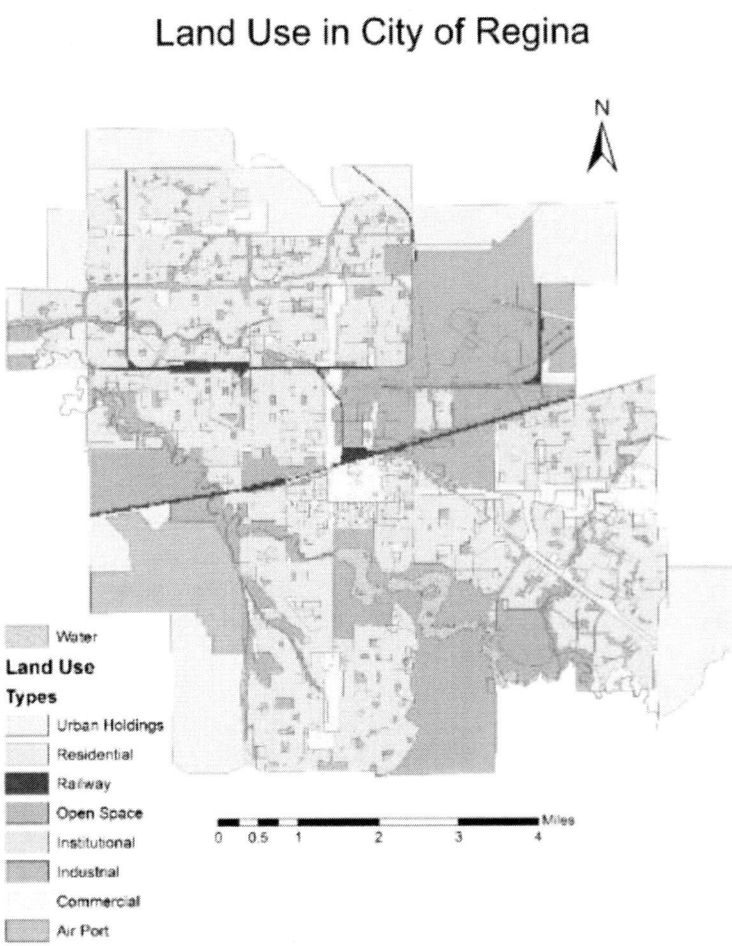

Figure 3: Current land use pattern in Regina.

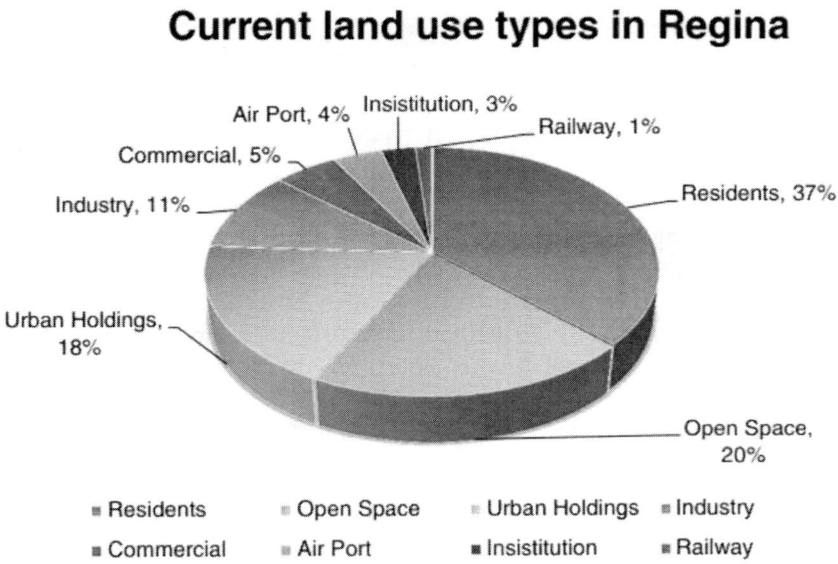

Figure 4: The current land use types and proportion in Regina.

GIS

GIS is a set of tools for inputs, storage and tetrieval, manipulation and analysis, as well as outputs of spatial data (Malczewski, 1999). ArcGIS is acknowledged to be a powerful tool in solving the spatial problems. ArcGIS by ESRI GIS and mapping software was applied for spatial data analysis and mapping in this study. All the related data were collected from Terrasever and Terra lab at University of Regina. Land-use maps and administrative information were input into GIS digitally to establish a new geo-database, then overlapped with each other.

Meanwhile, policy analysis based on community plans and literature reviews were completed, serving as foundation for land use type categories. According to available data, land use for human activity were divided into five suitability levels. In this process, multi-criteria analysis method was used for classifying and weighing criteria. Quantitative analysis is necessary for multi-criteria analysis, including scoring, ranking and weighting.

Finally, an output map of the land use suitability with five classes was displayed and a comparison was conducted between the new land use pattern and the pre-existing land use status. The Halme et al. approach introduces the decision-maker's preference in the efficiency analysis, by explicitly locating his most preferred solution vector on the efficient frontier.

The same authors highlight that when systematically exploring the neighborhoods of the Most Preferred Solution (MPS), one does not know explicitly the decision-maker's value function, but its form becomes known when the end of the search for MPS is reached.

Weight product

Weight product (WP) method has been introduced centuries ago and been advocated in the past few years. WP is a relative simple multiple attribute utility methods. Since WP is easily understood by decision makers and is easy to be conducted, this method have been widely applied in many fields. In this study, based on the WP method, three factors including population, employment and average income, hold significance in municipal land use planning. In the multi-criteria analysis process as showed in Figure 5, they were assign different weight to obtain a total score of every region based on the following formula (Gomes and Lins 2002):

$$S_f = \sum_{j=1}^{3} W_j * A_i \tag{1}$$

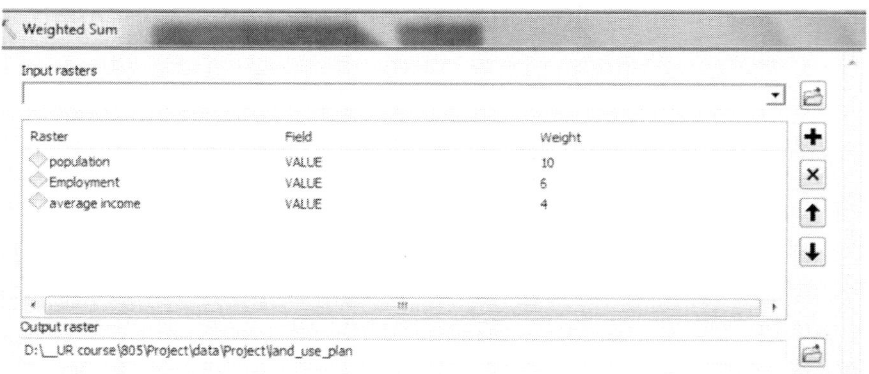

Figure 5: The Multi-criteria Analysis Process.

Where W_j --- is weight of each criterion j = 1,2,....m,

And A_i---is the normalize the value of each grid cell, i = 1,....., n.

The general MCAD approach for this case may be seen in (1), where X_i, Y_i,..., represent the value of the criterion X, Y..., for the alternation i; λ are the decision variables that represent the decision-maker's preferences for the alternative i, i = 1,...,n. For this case study, λ vector representing the decision maker preferences.

$$max\sum_{i=1}^{n} X_i * \lambda_i$$
$$max\sum_{i=1}^{n} Y_i * \lambda_i$$
$$\vdots$$
$$s.\ t.\sum_{i=1}^{n} \lambda_i = 1$$
$$\lambda \geq 0,\ \ i = 1, \ldots, n.$$

Defining the criteria

The expansion of land has overwhelmingly been a response to fast-rising population decades ago, so population is considered the most essential drive force of land exploitation. The weight of each criterion has been shown in Table 1. Meanwhile, as the increasing of urban population, the urbanization of Regina fringe is an inevitable trend. Study of population can not only help analysis the existing land use pattern, but also assist land use trend forecasting. Based on the above-mentioned consideration, that population is specified the maximum weight among the criteria. The population distribution of Regina is shown in Figure 6.The employment is expanded to labor force, which is in turn expanded to population equivalents. From a land use point of view, although a city or a region is usually studied as a whole, it is also necessary to examine employment changes brought about by changes in economic. Therefore, employment is defined as a factor in the process of land use or land development. The employment situation of Regina is showed in Figure 7.

Table 1: The weight of each criterion

Criteria	**Wj**
Population	10
Employment	6
Average_income	4

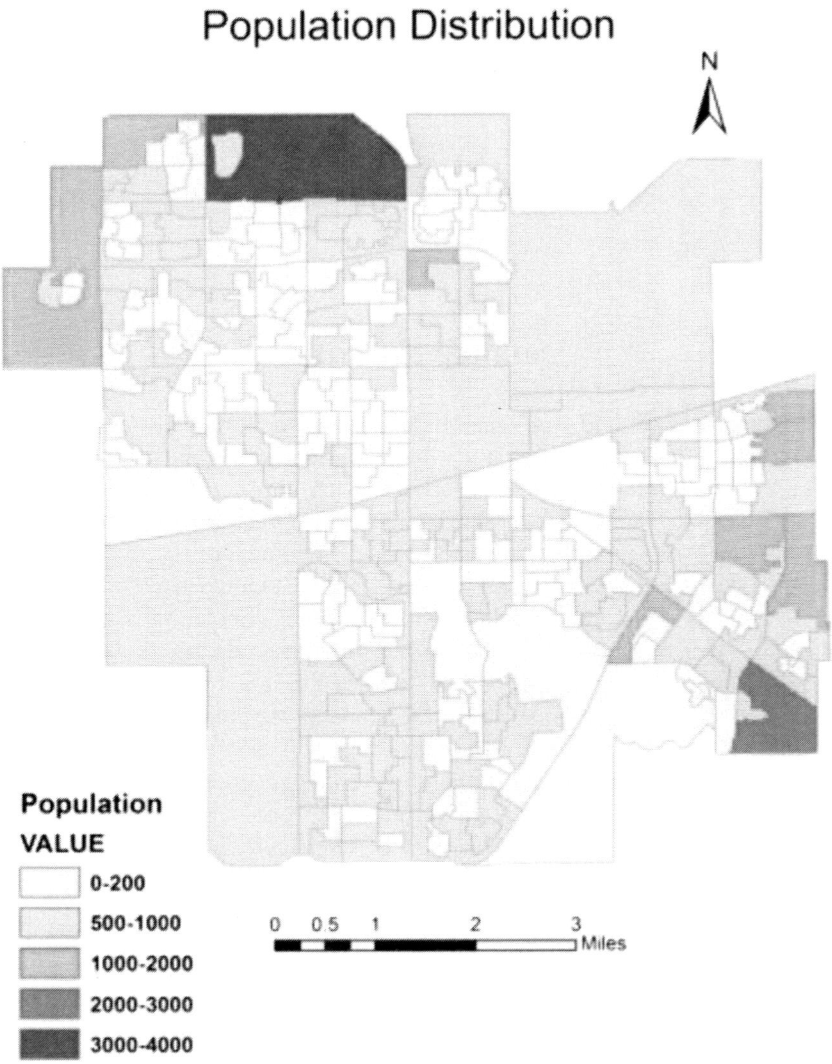

Figure 6: The population distribution of Regina.

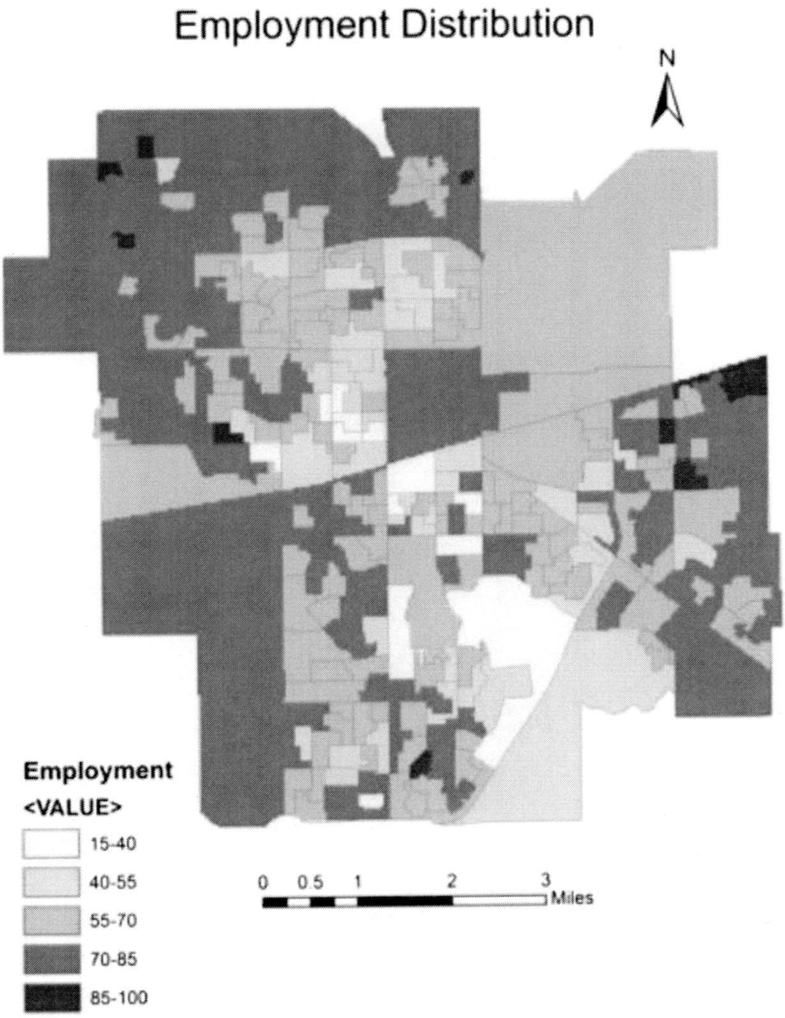

Figure 7: The employment situation of Regina.

Furthermore, as like Hok Lin Leung mentioned that income investigate is essential in house marketing, income is equivalent important in land use planning. The income census of a certain area can reflect the consumption level and trend in a certain extent (Leung 2003). Thus, as one of the economic factor, income is also taken into account when it comes to the land use suitability assessment. The average income of a certain census tract in Regina is showed in Figure 8.

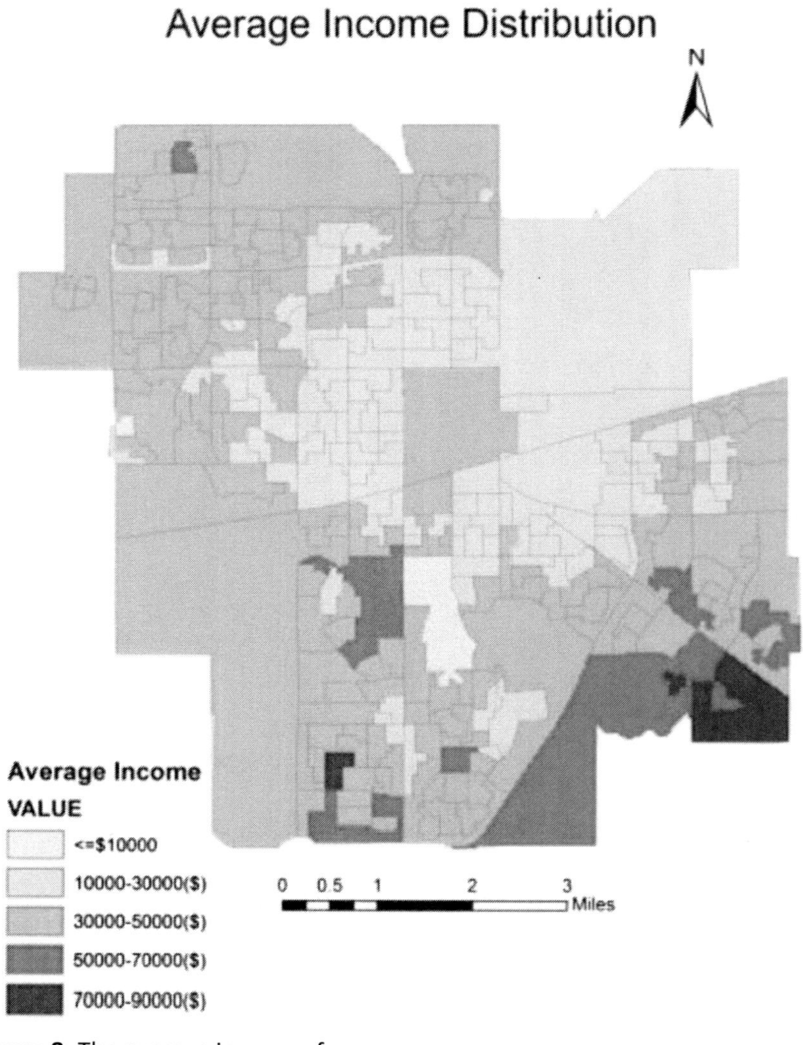

Figure 8: The average income of census.

RESULTS AND DISCUSSION

After the spatial analysis through ArcGIS, a land use suitable level map was obtained. In this map, the suitability of land use was divided into 5 classes (shown in Figure 9). The most suitable level was defined as the highest score among the weighted values, marking red in the legend. The red areas are in high rate of employment, as well as high income with large population, so they are capable for more human activities area like commercial and residential regions. On the contrary, the least suitable areas were defined to be more suitable for open sources, since there are less human beings and low income or benefit. Based on this interpretation, it could be predicted that the development tendency of Regina would be at north-west and south-east part of the city.

Figure 9: The land use suitability assessment output.

Meanwhile, the current land use map is jointed into the suitability assessment output (Figure 10). Comparison is conducted between these two land use patterns in Table 2. As shown in Table 2, there is a significant distinction between the simulated land use and the actual land use pattern. The results shows that there are significant difference between the simulated land use availability and the existing land use conditions. Although we can say it is not an ideal plan, we can still find and learn something faulty from the process of analysis and make it available to improve in the next time.

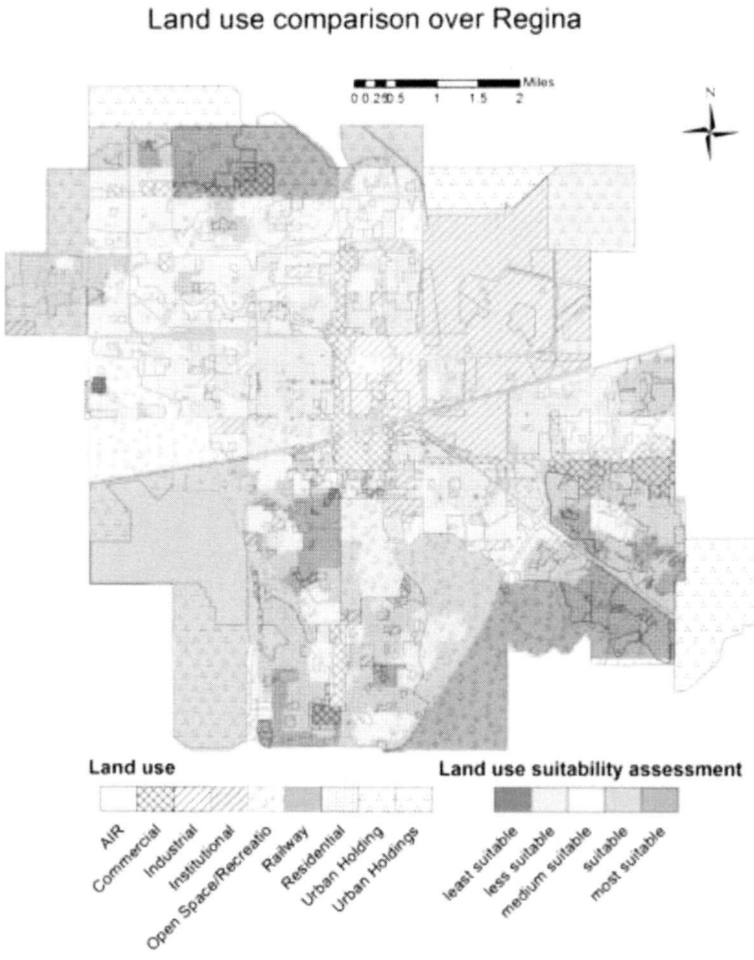

Figure 10: The comparison between current land use pattern and simulation.

In the study, the priority is the criteria selection and determination. In real-world case study, a large number of criteria should be determined, referring to environment, ecology, social and economic. Whilst in this research, only economic factor were considered. This can be considered as a single stage test of the complicated land use plan procedure. Moreover, the weight of criteria should be determined by both experts and stakeholders. In this research, the weight was just determined by literal evaluation from municipal land use project plan.

Table 2: The comparison between simulation and actual land use

Suitable level	Simulation land use	Actual land use
Least suitable	Urban holdings	A black spot
Less suitable	Air	Industry
	Open space/Recreation	Residential
Medium suitable	Industry	Residential
	Railway	Commercial
		Open space/Recreation
Suitable	Commercial	Air
	Institutional	Urban holdings
Most suitable	Residential	Commercial
		Urban holdings

Secondly, the simulation of suitable level is simplified, whilst in the real world, the development pattern is complicated. In the analysis, development scenarios in land use are not easy to pin down. There are no universal rules as to what they should or should not include, and there are no tools available to generate them automatically. F.C. Dai et al. set two scenarios in their study of urban fringe land use planning. One is mainly based on the actual development simulation, and the other scenario is mainly depended on the government planning. Thus, setting scenarios might increase the accuracy of the simulation as well as the reliability.

Moreover, as it is under a small scale analysis, the simplified definition is considered to be acceptable. However, the definition of suitability level results in unprofessional. Take the black spot in the least suitable level for example. It is hard to explain its appearance. It might be assumed to be the fault in the multi-criteria analysis procedure. And beyond the public opinion survey and social investigation, the suitability assessment seems to be groundless. Thus, the result might be better if the operational research is implemented in the future study.

CONCLUSION

Assessing the land use development problem has become an important task under the increasing aware of land resource conservation. Land use suitability assessment is a practical tool to make decision on land use development. However, less works could be found that use quantized index to evaluate the land resource usage at Prairie Provinces in Canada. This study applied GIS combined with multi-criteria decision analysis ideology in the City of Regina to assess land use condition with three important criteria referring to social and economic. Finally, a map with five suitability levels can be obtained.

When comparing the simulated land use suitable classes with the existing land use pattern, we can find that: the significant differences between these two land use patterns reflect the limitation of the approach and data availability. Two possibilities are put forward to increase the reliability and accuracy of land use suitability assessment, where the research is potentially valuable. Urban land use categories are complicated process. Recommendations made for future studies is to improve the efficacy and objectivity of local land use evaluation to support the land use suitability assessment and avoid the subjectivity. As for the accuracy of the real-world case study, the criteria selection and the weight assignment should be widely and deeply discussed and researched.

The results represent the potential of GIS-based evaluation for urban planning purpose. However, it needs to be emphasized that the reliability of the assessment results depends on a multitude of factors ranging from the quality of the database to the potential errors in the GIS. Meanwhile, the modeling results are highly significant to the weights applied. The determination of a multitude of factors and weights for the various factors is one of the most important challenges in the future.

AUTHOR'S CONTRIBUTIONS

JC carried out the data collection, methods, results analysis and finish the manuscript.

ACKNOWLEDGEMENT

This research has been conducted within University of Regina. The Terrasever and Terra lab provided the equipment and materials for this research. This research is also supported by Dr. Joe Piwowar in the data collection and Dr. Gordon Huang.

COMPETING INTERESTS

The author declares that she has no competing interests with anyone.

REFERENCES

1. Antoine J, Fischer G, Makowski M: **Multiple criteria land use analysis.** *Appl Math Comput* 1997, **83:** 195–215.
2. Cheng S, Chan CW, Huang GH: **An integrated multi-criteria decision analysis and inexact mixed integer linear programming approach for solid waste management.** *Eng Appl Artif Intel* 2003,**16**(5):543–554.
3. Collins M, Steiner F, Rushman M: **Land-use suitability analysis in the United States: historical development and promising technological achievements.** *J Environ Manage* 2001,**28**(5):611–621.
4. Dai F, Lee C, Zhang X: **GIS-based geo-environmental evaluation for urban land-use planning: a case study.** *Eng Geol* 2001,**61**(4):257–271.
5. Gomes E, Lins M: **Intergrating geographical information systems and multi-criteria methods: a case study.** *Ann Oper Res* 2002, **116:** 243–269.
6. Huang G, Xia J: **Barriers to sustainable water-quality management.** *J Environ Manage* 2001, **61:** 1–23.
7. Joerin F, Thériault M, Musy A: **Using GIS and outranking multicriteria analysis for land-use suitability assessment.***Int J Geogr Inf Sci* 2001,**15**(2):153–174.
8. Kiker G, Bridges T, Varghese A, Seager T, Linkov I: **Application of multicriteria decision analysis in environmental decision making.** *Integr Enviro Assess Manage* 2005,**1**(2):95–108.
9. Kunwar P, Kachhwaha T, Kuma A, Agrawal AK, Singh A, Mendiratta N: **Use of high-resolution IKONOS data and GIS technique for**

transformation of landuse/landcover for sustainable development. *Curr Sci India* 2010,**98**(2):204–212.

10. Leung H: *Land use Planning Made Plain*. University of Toronto Press, Toronto, Canada; 2003.

11. Liu Y, Lv X, Qin X, Guo H, Yu Y, Wang J, Mao G: **An integrated GIS-based analysis system for land-use management of lake areas in urban fringe.** *Land Scape Urban Plan* 2007,**82**(4):233–246.

12. Malczewski J: *GIS and multicriteria decision analysis*. John Wiley & Sons; 1999.

13. Malczewski J: **GIS-based land-use suitability analysis: a critical overview.** *Pro Plann* 2004,**62**(1):3–65.

14. Malczewski J: **GIS–based multicriteria decision analysis: a survey of the literature.** *Int J Geogr Inf Sci* 2006,**20**(7):703–726.

15. Phua M, Minowa M: **A GIS-based multi-criteria decision making approach to forest conservation planning at a landscape scale: a case study in the Kinabalu Area, Sabah, Malaysia.** *Land Scape Urban Plan* 2005,**71**(2):207–222.

16. Sharifi M, Boerboom L, Shamsudin K, Veeramuthu L: *Spatial Multiple Criteria Decision Analysis in Integrated Planning for Public Transport and Land use Development Study in Klang Valley, Malaysia*. ISPRS Technical Commission 2nd Symposium, United States; 2006:85–91.

2

RENEWABLES IN RESIDENTIAL DEVELOPMENT: AN INTEGRATED GIS-BASED MULTICRITERIA APPROACH FOR DECENTRALIZED MICRO-RENEWABLE ENERGY PRODUCTION IN NEW SETTLEMENT DEVELOPMENT: A CASE STUDY OF THE EASTERN METROPOLITAN AREA OF CAGLIARI, SARDINIA, ITALY

Claudia Palmas[1*], Emanuela Abis[2], Christina von Haaren[1] and Andrew Lovett[3]

[1] Department of Environmental Planning, Leibniz University of Hannover, Hannover 30419, Germany.

[2] Department of Civil, Environmental Engineering and Architecture, Cagliari University, Cagliari 09124, Italy.

[3] School of Environmental Sciences, University of East Anglia, Norwich NR4 7TJ, UK.

ABSTRACT

Background

In recent years, there has been an increasing interest in using micro-renewable energy sources. However, planning has not yet developed methodological approaches (1) for spatially optimizing residential development according to the different renewable energy potentials and (2) for integrating objectives of optimized energy efficiency with other environmental requirements and concerns. This study addresses these topics by firstly presenting a new concept for the regional planning.

Methods

The methodological approach for the evaluation of spatial variations in the available energy potential was based on the combination of existing methods adapted to the local scale and data availability. For assessing the bioenergy potential, a new method was developed. Other environmental criteria for deciding about sustainable locations were identified through a survey of more than 100 expert respondents. This survey involved pairwise comparisons of relevant factors, which were then translated into relative weights using the Analytical Hierarchy Process. Subsequently, these weights were applied to factor maps in a Geographical Information System using a weighted linear combination method.

Results

In the test region, the eastern metropolitan area of Cagliari, Sardinia, this analysis resulted in the designation of suitable areas for new settlements and preferred locations for micro-renewable technologies. Based on expert preferences, a number of alternatives for future housing development were identified, which can be integrated in the early stages of land use or development plans.

Conclusions

The method proposed can be an effective tool for planners to assess changes and to identify the best solution in terms of sustainable development.

Keywords: Micro-renewable energy potentials Multicriteria analysis Residential development Analytical Hierarchy Process Pairwise comparison methods Weighted linear combination

BACKGROUND

A sustainable future for cities significantly depends upon the integration of energy efficiency in regional and urban planning. About 40% of the final energy demand is needed to heat and power homes. This represents a major source of greenhouse gas emissions, making energy savings in the field of residential development a key element of the European climate change strategy [1]. In this context, Europe is also faced with the challenge of implementing growing amounts of intermittent power sources such as micro-solar and wind sources in the electricity grid. The generation of renewable energy is characterized by intermittency; therefore, it is imperative that a mix of sources should be selected and used along with the suitable energy storage mechanisms in order to best utilize the available renewable energy resources and ensure the continuity of supply [2].

Two European projects exemplify the state of the art in the energy-efficient residential development: the BedZED development in the south of London and the Vauban development in Freiburg, Germany. These two eco-districts are globally recognized to be models for sustainable environmentally oriented planning using solar energy (photovoltaics and solar thermal collectors). In the BedZED project, the use of solar energy is maximized through the integration of solar cells into the vertical south-facing facades and also through a large installation on the south-facing roofs [3].

In Freiburg, the principles of energy savings and solar optimization are early considered in the planning phase of housing development, e.g., by defining the orientation and position of buildings or by obligatory low-energy construction requirements [4]. These urban multi-residential housing developments are not only models for energy saving, but they also take into account social and economical aspects.

However, the geographical distribution of the renewable energy potential is rarely considered or estimated in the planning of new residential areas [5, 6]. Also, in selecting the location, environmental criteria in combination with micro-renewable potentials are still neglected.

Therefore, the state of the art in the field of energy efficiency [7, 8] may be advanced by combining the different energy sources in new housing developments and planning their location by estimating the energy potential available for the whole area under consideration. Renewable energy supply is site-specific and variable [9]. A restriction on secure supplies from single renewable energy sources is their output variability. Supply insecurity can be increased by demand variability, especially where this correlates with times of high energy output by renewables, better predictability of their generation output, and the complementarities of different power sources. Also, the combination of different renewable energy sources can increase the supply security.

In recent years, several renewable energy potential mapping methodologies have been developed (e.g., solar irradiation and wind estimation, geothermal and biomass energy) [10, 11, 12]. These methods can be used for complementing the urban planning approaches. However, the methodologies have been developed for very small scales and cannot be applied unmodified for selecting new housing locations [13]. Therefore, it is necessary to either adapt the existing methodologies or develop new ones.

Energy efficiency should be integrated from the start of the land use planning process in order to guide the future development to support the sites with the best potential for using renewable micro-generation. These potentials can be developed in a sustainable way by using multicriteria evaluation methods in a Geographical Information System (GIS) to help optimize new settlements in terms of multi-functionality. There is a history of research using such multicriteria evaluation techniques to support collaborative decision-making processes by providing a framework where stakeholder groups can explore, understand, and redefine decision problems with respect to housing location [14, 15].

The development and testing of a methodology for an integrated approach to energy-efficient residential development planning is the main objective of the research presented here. Both optimizing the location of new housing development with regard to energy supply and other sustainability criteria as well as optimizing the mix of micro-renewables need to be facilitated. Consequently, the main research questions addressed are

- How to calculate the geographic distribution of energy potentials? How to produce energy potential maps? How to identify the best energy mix combinations? Which criteria and algorithms are needed for identifying the theoretical energy potential in terms of the different energy sources?
- Which environmental and landscape criteria are considered most relevant for the assessment of new housing development with micro-renewable technologies?
- How to support decision makers or planning in the challenge of including multiple criteria in housing development decisions?

Accordingly, this paper describes

- Methodologies (existing, adapted, or newly developed) to estimate the micro-renewable energy potentials in a spatially explicit manner.
- Methodologies for identifying the suitable areas for new sustainable settlements using micro-renewable technologies which enable to support decision makers in planning.

The results of testing these methods are presented for the region of Cagliari in Sardinia.

METHODS

General methodological approach

Part of the approach is not only based on the existing, originally small scale, but also on methods for energy potential assessment, which were pre-tested in the Hannover region (by Master students in cooperation with the State Office for Mining, Energy and Geology (LBEG) [16]. In a second step, the methods were adapted to local/regional scale planning. As there was no suitable method for assessing the bioenergy potential, a new method was developed. This resulted in assessments of the theoretical (potential) supply. Because of the existing technical, ecological, economic, and social restrictions, such theoretical amounts can only be exploited up to a certain percentage [17].

Expert preferences were used to weight multiple assessment criteria for housing developments. These preferences were obtained through a survey conducted with students, academic planners, regional planners, and public authorities in Italy, Germany, and the UK. This expert-based approach was chosen because in most European countries, no clear-cut standards exist about the suitability of micro-energy generation in residential areas (in contrast, e.g., to emission standards). Expert opinions were a simple way of priority setting in such complex decisions. In addition, such a method allows the results of different preferences to be modeled and local or regional stakeholder opinions and interests to be included.

The energy potentials and expert preferences were ultimately combined in a GIS-based analysis to identify the most appropriate housing sites on a regional scale. This analysis made use of multicriteria evaluation (MCE) techniques which are one of the most common GIS-based tools. They have been used to support decision making on complex problems such as site selection, land suitability, resource evaluation, and land allocation [18, 19, 20, 21, 22]. Over the last two decades, several MCE methods have been implemented in a GIS environment, including weighted linear combination (WLC) and its variants [23, 24, 25, 26, 27] and ideal point methods [28, 29]. Among these procedures, WLC and Boolean overlay operations, such as intersection (AND) and union (OR), are the most widely used [15, 21] and were adopted in this research.

Data

The eastern metropolitan area of Cagliari (Figure 1) covers 591 km^2 in the south of Sardinia and has a population of 322,392 inhabitants [30]. Cagliari is the capital of Sardinia, situated at the southern shore of the island and has 157,222 inhabitants [30].

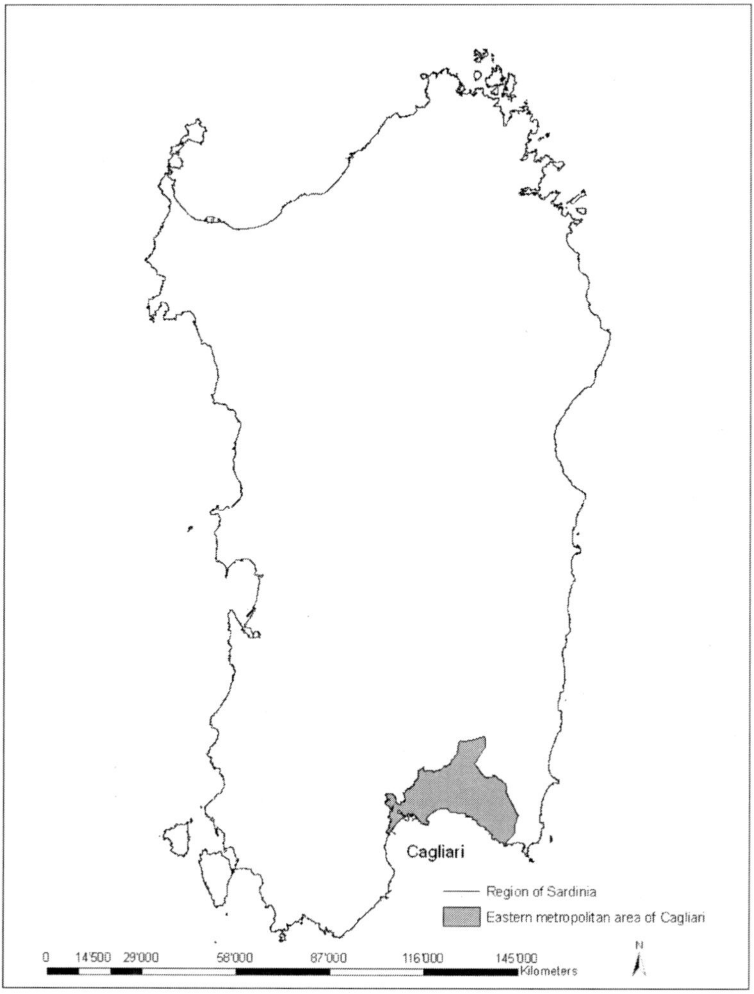

Figure 1: Eastern metropolitan area of Cagliari.

The region is characterized by rural areas around the cities with a large amount of agricultural land (around 46.72%). Other uses, such as residential, commercial, and industrial areas, cover about 40% (land use data, region of Sardinia).

Table 1 lists the main geographical data sources used for the regional assessment. These were supplemented by shape files from the Regional Landscape Plan of Sardinia, scale 1:10,000.

Table 1: Input data for the energy potential estimation

Data	Scale/unit	Data origin
Digital elevation model (DEM 90)	90 × 90 m	CGIAR Consortium for Spatial Information
Wind speeds at 25 m	m/s	Aeolic Italian Atlas
Geological map	1:200,000	Earth Science Department (Cagliari University)
Land use	1:25,000	Region of Sardinia
Irrigation map	1:25,000	Region of Sardinia

Adaptation of an existing methodology for identifying the spatial solar energy potential under local-scale conditions

The solar potential raster maps were calculated from the *r.sun* model. The *pvgis* database, derived from the Photovoltaic Geographical Information System - Interactive Maps [31], was used to validate the data. The algorithm used to calculate the solar irradiation was implemented in the open-source GIS software GRASS, where the beam irradiance normal to the solar beam B_{0c} (in watts per square meter) is attenuated by cloudiness atmosphere and calculated in the model as in Equation 1 [31]:

$$B_{0c} = G_0 \exp\{-0.8662\ T_{LK}\ m\ d_R(m)\}, \tag{1}$$

where G_0 is the extraterrestrial irradiance normal to the solar beam (in watts per square meter), −0.8662 T_{LK} is the Linke atmospheric turbidity factor, m is the 'optical air mass,' and $d_R(m)$ is the 'Rayleigh optical thickness at air mass m.'

The *r.sun* model operates in two modes. In *mode 1*, the model calculates the instant time (in seconds) of raster maps of chosen components (beam, diffuse, and reflected) of solar irradiance (in watts per square meter) and the solar incident angle (in degrees). In *mode 2*, the raster maps of the daily sum of solar irradiation (in watt hours per square meter per day) are computed as an integration of irradiance values that are calculated within a set day. In this study, *mode* 2 was used because we needed to calculate raster maps representing the annual average of daily sums of global irradiation for horizontal surfaces. To compute the irradiation raster maps, *r.sun* requires only a few mandatory input parameters - digital terrain model (elevation, slope, aspect - *elevin slopein aspin*), *day* number (for mode 2), and additionally, a local solar *time* (for mode 1). The other input parameters are either internally computed (solar declination) or the values can be set to fit the specific user needs: the Linke atmospheric turbidity, ground albedo, beam, and diffuse components of clear-sky index, and time step are used for the calculation of all-day irradiation [32].

Adaptation of an existing methodology for identifying the spatial wind energy potential

The wind speeds were calculated in accordance with the following Equation 2 [[33, 34]]:

$$v = v_{\mathrm{ref}} \left(\frac{z}{z_{\mathrm{ref}}} \right)^{\alpha}, \tag{2}$$

where v = wind speed at height z above the ground level; v_{ref} = reference speed, i.e., a wind speed we already know at height z_{ref}; z = height above the ground level for the desired velocity, v; and z_{ref} = reference height, i.e., the height where the wind speed is measured v_{ref}.

The exponent α is an empirically derived coefficient that varies according to the stability of the atmosphere. For neutral stability conditions, α is approximately 0.143.

Adaptation of an existing methodology for identifying the spatial geothermal energy potential

The geothermal energy potential maps were generated by considering the physical rock properties for the estimation of the specific heat extraction values, where those for vertical loops followed Kaltschmitt et al. [35]. Here, the geological stratification of rocks to 100 m is derived from a regional geological map and the specific heat extraction is obtained from the following equation:

$$P_{\mathrm{EWS}} = (13 \cdot \lambda) + 10, \tag{3}$$

where P_{EWS} = specific heat extraction capacity, and λ = heat conductivity of the rock.

To obtain the specific heat extraction values, the geology was divided in two homogeneous layers: unconsolidated and solid rocks. Further information on the geological stratification for vertical loops and soil characteristics for horizontal loops was obtained from Dott. Geol. Fausto Pani, freelance, and Prof. Giovanni Barrocu, Cagliari University.

Developing a methodology for identifying the spatial biomass energy potential

Given the focus on housing development, not every type of biomass is relevant. Attention was focused on wooden biomass which is suitable for producing heat and electricity in residential areas with the installation of a cogeneration system. Important criteria for identifying the potential include the distance of the source of wood from the settlement and the capacity of the forest in terms of the wood reservoir. From an economic perspective, the energy-efficient use of biomass can be defined as a use within a radius of 30 km around a potential biomass facility [36]. According to the sustainability principles, e.g., the needs of localizing new settlements near the biomass source, we assume that the energy biomass efficiency is related to a use within a radius of 15 km around a potential biomass facility, as shown in Equation 4. We assume that the biomass energy potential P_i is defined as the theoretical qualitative potential for a hypothetical settlement location or users V_i.

$$P_i := \sum_j \left[\left[\frac{A_j}{A} \cdot \frac{(15 - d_{ij})}{15} \right] - T_j \right] * i \qquad (4)$$
$$= 1, 2, K, \mathrm{N}; j \neq i$$

where P_i = biomass energy potential, V_i = potential settlement, A_j is the forest cell area, A is the total forest cell area, d_{ij} is the distance between the centers of the cell of settlement potential location and of the cell of the forest areas $d_{ij} \leq 15$ km, and T_j is the factor depending on transport and wood extraction cost.

To differentiate between the areas of varying potential, a Monte Carlo method was introduced. Broadly speaking, Monte Carlo integration methods are algorithms for the approximate calculation of the numerical value of a definite integral, usually multidimensional ones, in our case the sum of forest areas (*cf.*[37]). The usual algorithms evaluate the integrand at a regular grid. Monte Carlo methods, however, use random samplings to approximate probability distributions. This is performed by selecting some numbers of random points over the desired interval and summing the function evaluations at these points [38].

Best energy mix combinations

The suitability maps of the theoretical energy potentials were integrated into combined layers showing the best locations for the new settlement development according to the most appropriate energy mix for the area under consideration. Maps were normalized and two versions were produced depending on whether geothermal vertical or horizontal loops were included.

Survey of expert preferences

Decisions about the spatial resource allocation require prioritizing multiple criteria. Different criteria were selected for (1) assessing housing development in general as well as for (2) settlements with micro-renewables. The selection of criteria took into account the possible environmental and landscape impacts as well as the availability of relevant geodata in order to transform the preferences into spatially explicit representations. The criteria used in this research were proximity to existing urban areas, proximity to major roads and train lines, distance from environmentally valuable and vulnerable areas or from protected areas, proximity to water (lakes and rivers), and the slope gradient (see Table 2).

Table 2: Criteria for new settlement development

Factor/criteria	**Type**
Proximity to existing urban areas	Planning factor (compact development)
Proximity to major roads and train lines	Transport factor
Distance from environmentally valuable and vulnerable areas or from protected areas	Environmental factor
Proximity to water (lakes and rivers)	Attractiveness factor
Slope gradient	Physical factor

Other factors such as the location, size, and accessibility of a site and its proximity to amenities and services are also important for future housing developments. These could be considered on a broader scale.

The criteria were divided into continuous suitability factors and constraints (binary yes/no restrictions). The constraints were built-up areas, water (lakes and rivers), and areas characterized by hydrogeological instability.

The criteria for the survey of the settlements with micro-renewables shown in Table 3 focused on landscape and environmental impacts, because the technical factors were included in the potential maps.

Table 3: Criteria for settlement development using micro-renewable technologies

Micro-technology	Criteria/factors
Solar panel and thermal collectors (S)	Distance from landscape-protected areas and other beauty areas
	Distance from historic/cultural facilities (historical centre, areas of historical and cultural interests, archeological sites)
Wind turbines (W)	Distance from historic/cultural facilities
	Distance from Special Protection Areas (SPA) and others avifaunistic important areas
	Distance from landscape-protected areas or other beauty areas
Biomass power plants (B)	Distance from historic/cultural features
	Distance from landscape-protected areas or other beauty areas
Geothermal vertical loops (GVL)	Distance from historic/cultural features
	Distance from drinking water or aquifers
Geothermal horizontal loops	Distance from historic/cultural features

Micro-technology	Criteria/factors
(GHL)	
	Distance from flooding areas

Both parts of the survey were conducted in Italy, Germany, and the UK and sought to gain insights into perceptions about new energy-efficient settlement development. This required the participation of people who had expert knowledge regarding landscape and environmental planning and/or renewable energy, so the survey focused on students and academic planners, regional planners, and public authorities. The questionnaire was distributed in person and by email, with participants returning the completed surveys in the same ways.

The expert preferences were converted into values using pairwise comparison methods, a procedure in the Analytical Hierarchy Process (AHP) [39]. As an input, the method takes the pairwise comparisons of the different criteria and produces their relative weights as an output. According to the relative importance, the weights, which were assigned to the different criteria, were calculated using MathCAD, an engineering calculation software. Consistency ratios were also calculated to assess the reliability of the pairwise comparisons [39].

The output maps were generated using a Boolean approach and a WLC method [21]. The Boolean approach is based on a reclassification operation and specified cutoffs. WLC was used to produce suitability raster maps for housing development and micro-renewable preferences with respect to environmental and landscape impacts. The suitability maps were generated as shown in Equation 5:

$$\text{Suitability map} = \Sigma[\text{factor map(cn)} * \text{weight(w} * \text{constraint(b0/1)}], \tag{5}$$

where cn = standardized raster cell, wn = weight derived from AHP pairwise comparison, b0/1 = Boolean map with values 0 or 1, and n = number of raster cell

To identify the optimal sites for new residential areas by using micro-renewables, the three GIS layers (energy potentials, suitability for new settlement development, suitability for new settlements with micro-generators) were overlaid. This integration was conducted using Spatial Analyst functions available in ArcGIS 9.x [40].

RESULTS

Solar energy potential

Parameters such as the albedo factor (0.2) and the Linke turbidity (3.0) were assumed constant across the region as a first approximation. The clear-sky indices were not available. The influence of terrain shadowing was taken into account by setting the -s flag. After validation of the data, the output raster map showed the annual average of the daily sums of global irradiation for horizontal surfaces (in watt hours per square meter per day) (see Figure 2).

Figure 2: Annual average of the daily sums of global irradiation for horizontal surfaces (Wh/m ^{2} /day).

Wind energy potential

To create the wind energy potential maps, speeds at 25 m above the ground with 1-km resolution were used. The data were derived from the Italian Atlas Wind Energy (Atlante eolico italiano) developed by the Genoa University and the CESI research center [41]. Equation 2 was applied to obtain a final average wind speed raster map (at 10 m above the ground) with a resolution of 90 m (see Figure 3).

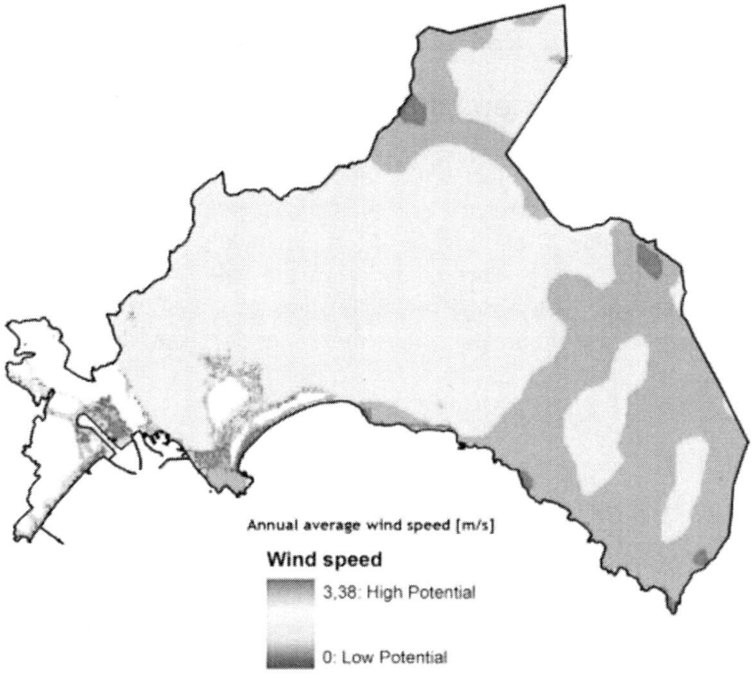

Figure 3: Wind energy potential.

Geothermal energy potential

For the unconsolidated rocks, there were some data limitations. Therefore, the thickness was sometimes only a rough estimate. The data for solid rocks were more accurate. The information regarding the groundwater flow component was not considered according to the VDI 3640 German directive.

Geothermal vertical loops

The geological map of the region of Sardinia, scale 1:200,000, was consulted to evaluate the specific heat extraction capacities, which were combined by values from the literature with regard to the specific heat conductivity (*cf.*[42]). The resulting map was classified into three categories (see Figure 4). The unsuitable areas are not suitable for economic reasons.

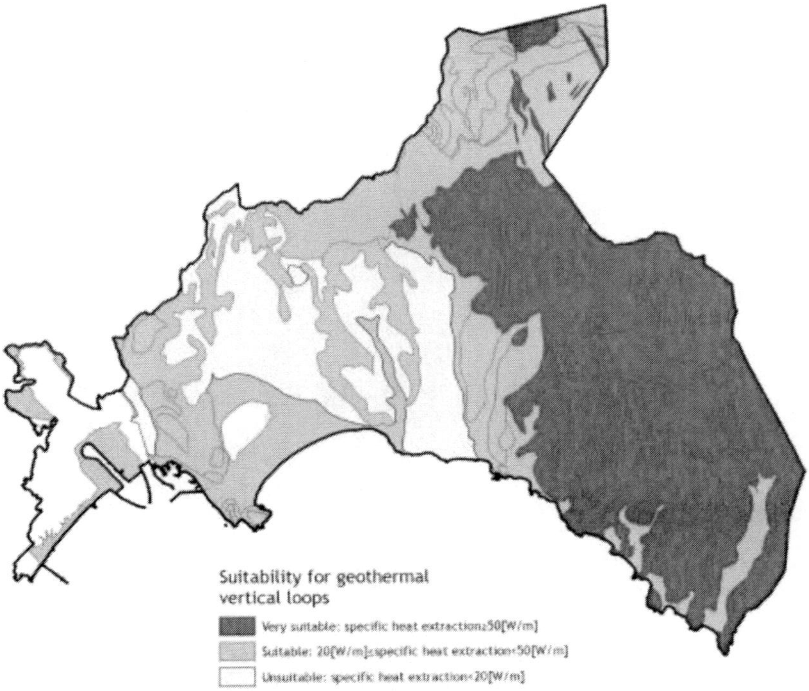

Figure 4: Geothermal energy potential for geothermal vertical loops.

Geothermal horizontal loops

The geological map, the map for irrigation, and the land use map (scale 1:25,000) were considered to select the suitable and unsuitable areas for the installation of horizontal loops. Given the variety of soil conditions (e.g., evapotranspiration) and characteristics (e.g., presence of aquifers), soil types, and the absence of quantitative data regarding all these factors [43], it was only possible to give a qualitative potential estimation for the use of horizontal loops on this scale, as is shown in Figure 5.

Biomass energy potential

The study did not take into account the factor T_j in Equation 4 which depends on road types and conditions as well as variable factors such as fuel prices for wood transportation and extraction costs, because the necessary data were not available. These factors can be better considered in a more detailed local view.

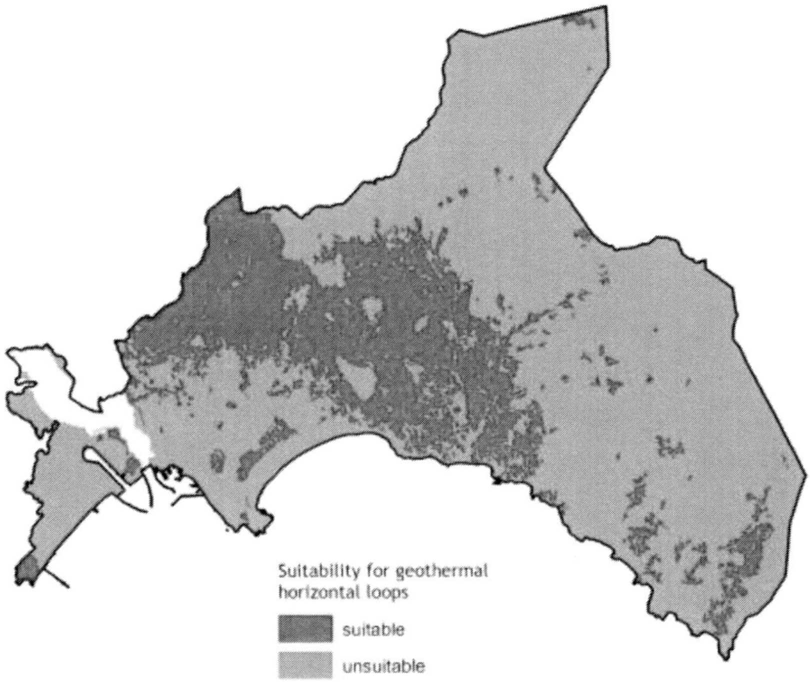

Figure 5: Geothermal energy potential for geothermal horizontal loops.

A grid with a 250-m spacing was overlaid over a larger section of the eastern metropolitan area of Cagliari. A total of 5,000 random points were used, giving an average density of around 1 point per 250 m^2. After the application of the Monte Carlo integration to the existing data, a biomass potential map was generated (Figure 6).

Best energy mix potential

We assumed that all the micro-energy potential maps are of the same weight. The maps obtained show the best locations for the integration of solar, wind, and biomass micro-generators with geothermal vertical loops (Figure 7) or horizontal loops (Figure 8).

Survey results

A total of 120 questionnaires were completed, but only 108 were considered further (consistency ratio < 0.1). As shown in Table 4, for each category of experts, we had a minimum of 15 valid questionnaires.

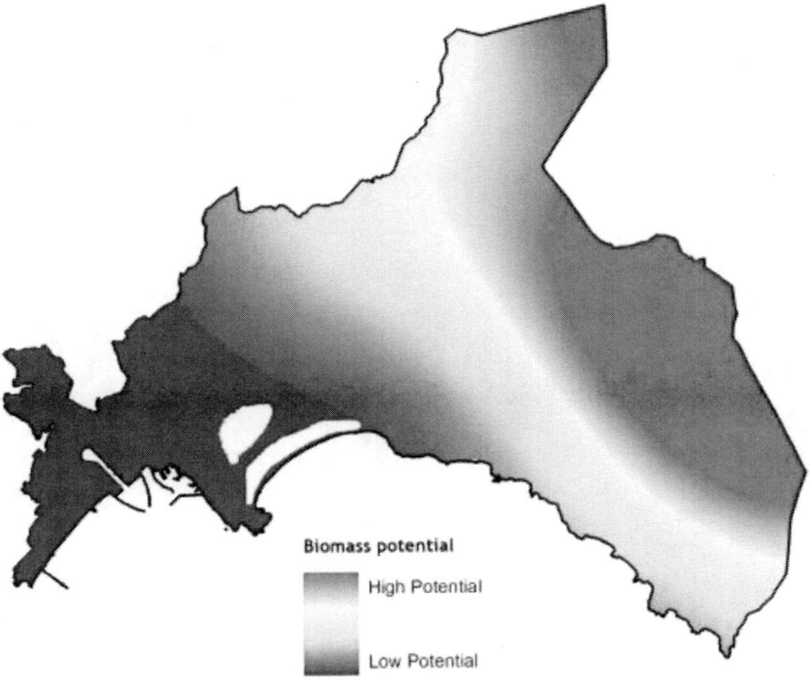

Figure 6: Biomass energy potential.

Obtained from students and academic planners (S.&AP.) as well as regional planners and public authorities (P.&PA.). DE, Germany; IT, Italy; UK, United Kingdom.

English experts preferred a compact development close to the built-up areas (S.&AP.: weight 0.29; P.&PA.: weight 0.35). The German experts gave the same weight to an urban development near roads and train lines (0.22). In terms of the distance from the environmentally valuable and vulnerable areas, Italian and German students as well as academic planners expressed a similar preference (0.34; 0.31) with the Italian and German regional planners and public authorities (0.23; 0.24). The Italian experts preferred a development that is close to lakes and rivers for attractiveness reasons (0.15; 0.17). German and Italian regional planners and public authorities gave the same consideration to the slope gradient (0.15).

Table 5 presents the weights for each criterion regarding the new housing development obtained from students and academic planners (S.&AP.) as well as regional planners and public authorities (P.&PA.) from each nationality. The weights sum to 1 with a higher value corresponding to more emphasis on the relevant criteria.

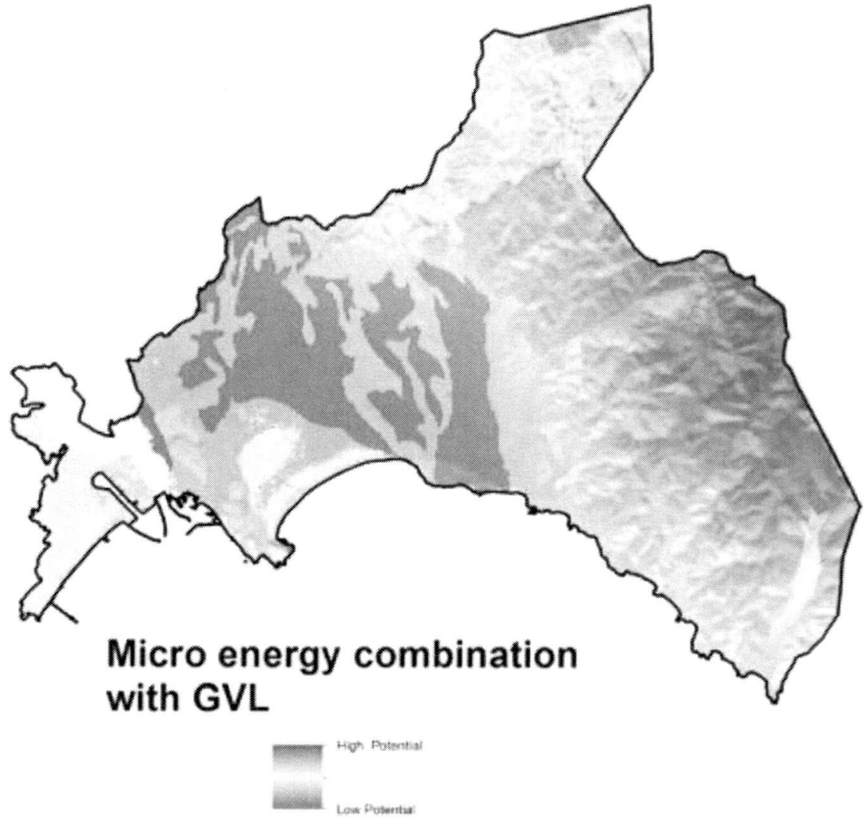

Figure 7: Best energy mix potential for solar, wind, and biomass micro-generators using geothermal vertical loops.

Table 4: Total evaluated questionnaires for each country

Experts	Total evaluated questionnaires with consistency ratio < 0.1		
	DE	**IT**	**UK**
S.&AP.	19	16	15
P.&PA.	15	28	15

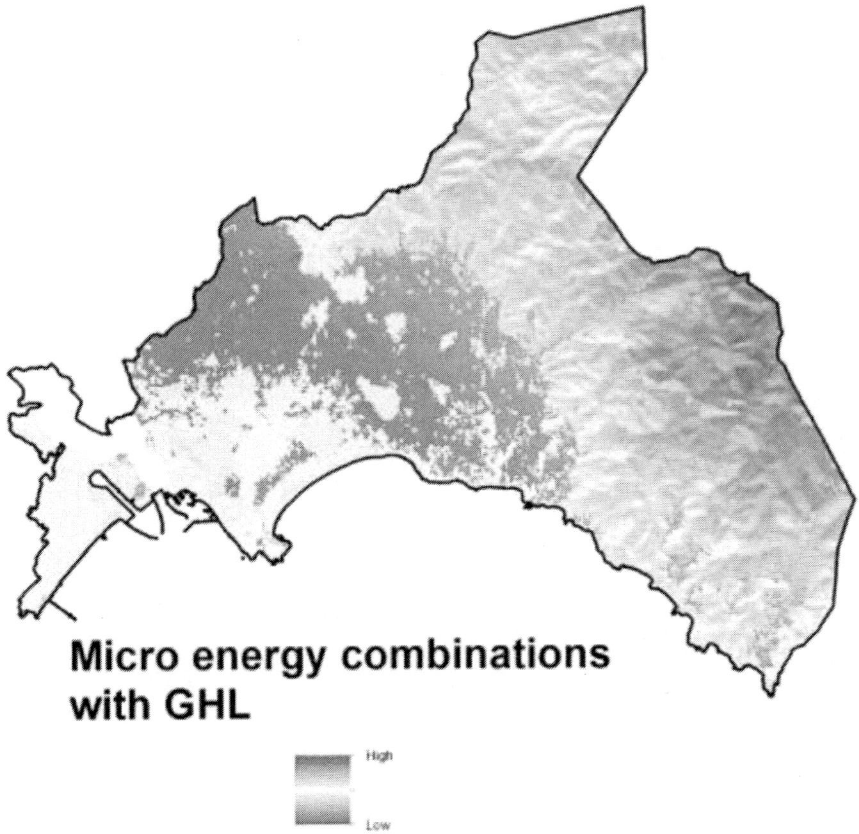

Figure 8: Best energy mix potential for solar, wind, and biomass micro-generators using geothermal horizontal loops.

Table 5: Weighting for settlement development

Criteria/factors	Experts	Weights		
		DE	IT	UK
Proximity to existing urban areas	S.&AP.	0.20	0.26	0.29
	P.&PA.	0.26	0.25	0.35
Proximity to major roads and train lines	S.&AP.	0.22	0.16	0.26
	P.&PA.	0.22	0.20	0.22

Criteria/factors	Experts	Weights		
		DE	IT	UK
Distance from environmentally valuable areas	S.&AP.	0.31	0.34	0.21
	P.&PA.	0.24	0.23	0.19
Proximity to water	S.&AP.	0.14	0.15	0.11
	P.&PA.	0.14	0.17	0.12
Slope gradient	S.&AP.	0.13	0.10	0.12
	P.&PA.	0.15	0.15	0.13

Obtained from students and academic planners (S.&AP.) as well as regional planners and public authorities (P.&PA.). DE, Germany; IT, Italy; UK, United Kingdom.

Table 6 shows the averages of the standard deviations expressed as a percentage of the means for the different expert groups and nationalities. These results show some variation in weighting but do not exceed 100% so the variations are not too high.

Table 6: Averages of standard deviations in percentages of the means for housing development

Experts	Average of the SD in percentage of the mean		
	DE	IT	UK
S.&AP.	65.74	79.62	49.78
P.&PA.	48.02	66.83	40.50

S.&AP., students and academic planners; *P.&PA.*, regional planners and public authorities; *SD*, standard deviation; *DE*, Germany; IT, Italy; *UK*, United Kingdom.

The final suitability maps were identified through the expert survey localizing the new settlement with renewable energy (see Table 7). The German students and academic planners (weight 0.54) as well as regional planners and public authorities (weight 0.60) gave more consideration to the visual impact caused by solar panels and solar thermal collectors on the cultural heritage. Italian academic (0.58) and environmental planners (0.54) and English academic (0.54) and environmental planners (0.54), by contrast, more intrusively considered the solar power plants near landscape-protected areas and other beautiful areas.

Table 7: Weighting for housing development using micro-renewable technologies

Criteria/factors	Experts	Weights		
		DE	IT	UK
Distance from landscape-protected areas and other beauty areas (S)	S.&AP.	0.46	0.58	0.54
	P.&PA.	0.40	0.54	0.54
Distance from historic/cultural facilities (S)	S.&AP.	0.54	0.43	0.46
	P.&PA.	0.60	0.46	0.46
Distance from historic/cultural facilities (W)	S.&AP.	0.26	0.35	0.30
	P.&PA.	0.31	0.40	0.35
Distance from Special Protection Areas (SPA) and avifaunistic important areas (W)	S.&AP.	0.50	0.45	0.39
	P.&PA.	0.44	0.30	0.38
Distance from landscape-protected areas or other beauty areas (W)	S.&AP.	0.25	0.19	0.31
	P.&PA.	0.25	0.30	0.28
Distance from historic/cultural features (B)	S.&AP.	0.30	0.38	0.34
	P.&PA.	0.35	0.47	0.39
Distance from landscape-protected areas or other beauty areas (B)	S.&AP.	0.70	0.62	0.66
	P.&PA.	0.65	0.53	0.61

Criteria/factors	Experts	Weights		
		DE	IT	UK
Distance from historic/cultural features (GVL)	S.&AP.	0.31	0.45	0.36
	P.&PA.	0.37	0.29	0.68
Proximity to drinking water or aquifers (GVL)	S.&AP.	0.69	0.55	0.64
	P.&PA.	0.63	0.71	0.32
Distance from historic/cultural features (GHL)	S.&AP.	0.43	0.48	0.39
	P.&PA.	0.42	0.54	0.53
Distance from flooding areas (GHL)	S.&AP.	0.57	0.52	0.61
	P.&PA.	0.38	0.46	0.47

GVL, geothermal vertical loops; *GHL*, geothermal horizontal loops; *S.&AP.*, students and academic planners; *P.&PA.*, regional planners and public authorities; *SD*, standard deviation; *DE*, Germany; *IT*, Italy; *UK*, United Kingdom.

German experts paid more attention to the environmental impact represented by the distance of wind turbines from important avifaunistic areas (respectively, weights: S.&AP., 0.50 and RP.&PA., 0.44). On the contrary, Italian experts expressed their preferences to the visual impact near historical and cultural facilities (S.&AP., 0.35; RP.&PA., 0.40), while English experts assigned almost equal weights to all three criteria, including the visual impact to landscape evaluable areas.

All experts, in particular the Italian regional planners and public authorities (0.71) as well as the German students and academic planners (0.70), assigned the highest weight to the criteria 'Distance from drinking water or aquifers' for geothermal vertical loops. Similarly, the experts, except for the English regional planners and public authorities (0.32), were in agreement regarding the importance of 'Distance from flooding areas' (average 0.63).

However, Italian experts assigned a similar weight regarding the visual impact of an additional chimney for a single power plant or a central biomass power plant near cultural/historical areas (0.48; 0.47) and landscape areas (0.52; 0.53).

The results of the survey showed similarities and differences between the stakeholder group preferences from the three countries. This outcome stems from national contrasts in planning systems and in attitudes towards micro-renewables. Transferring these preferences into a spatial representation resulted in an environmental suitability map which was overlaid with the energy potential for each micro-renewable technology. Figure 9 presents the three layers for solar energy. The combined result in Figure 10 shows the optimum sites for a new settlement development according to the energy potentials and expert preferences.

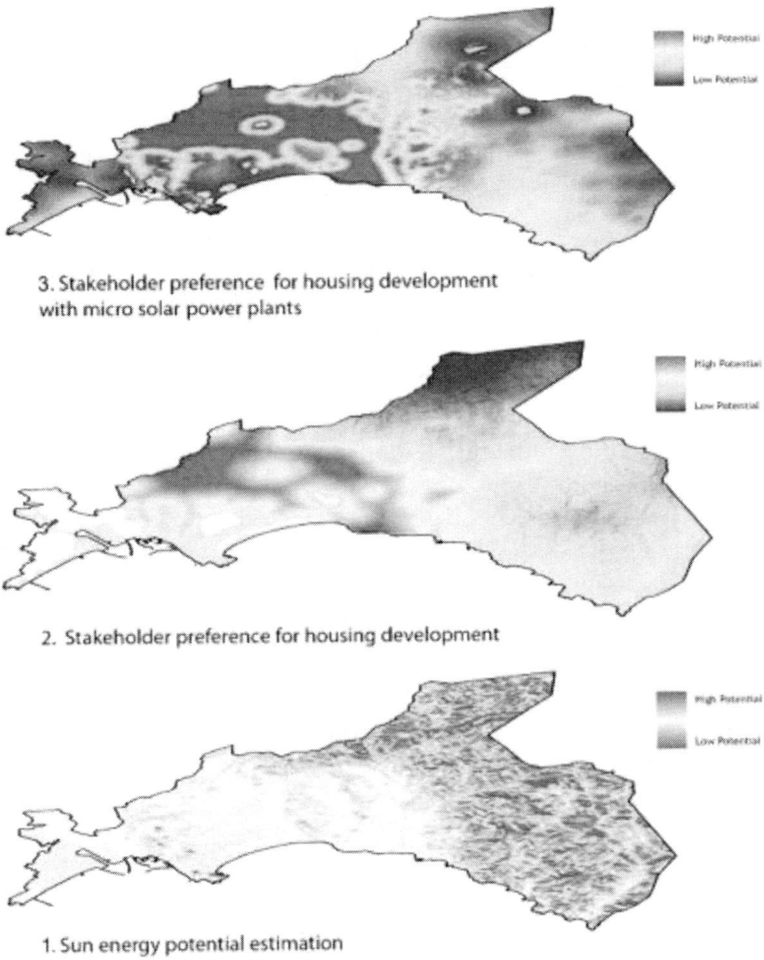

Figure 9: Overlaying of the three GIS layers (energy potential, settlement development, and settlement development using micro-generators).

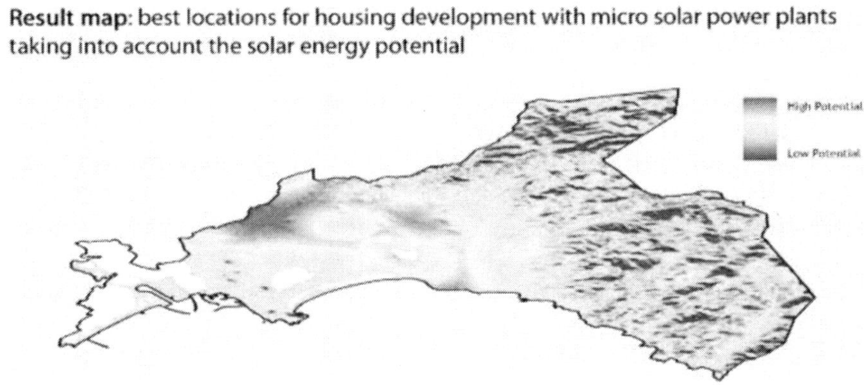

Figure 10: Suitability map for housing location using micro-solar power plants and energy potential.

The suitability maps resulting from the expert surveys were compared with the micro-energy potentials. There were many areas where the energy potential was high, which also corresponded to the expert preferences, where new settlements with renewable energies should be located.

The spatial results for solar irradiation reflect an (almost similar) high potential for the whole case study area. Nevertheless, areas where the potential is relatively low (areas in blue), because of the terrain aspect and slope, should be excluded. The wind potential varies along the Cagliari region.

With respect to the geothermal vertical loops, all experts assigned the highest weight to the criteria 'Distance from drinking water or aquifers.' Geothermal vertical loops should be buried up to 100 m deep and in some cases can modify the groundwater flow with consequences on the new settlements (*cf.*[44]), and on the water quality and temperature. This has to be taken into account in the planning of new urban settlements.

The geothermal energy potential map for geothermal horizontal loops and the suitability map of the expert groups showed no compliance. This is indicating that conflicts could arise if geothermally powered settlements will be planned. For this reason, it may be important to make decisions according to other criteria or needs.

Only few areas located in the east of the metropolitan area showed a good biomass potential and were at the same time suitable according to the expert preferences. Consequently, other energy sources should be chosen in most of the areas far away from any forests.

DISCUSSION

Decisions about the best energy mix for the different residential areas can be supported by the presented results. The proposed approach is based on an assessment of the energy potential and other relevant criteria which have been weighted by expert preferences.

A method has been developed for both the integrated assessment of four renewable energy sources and the identification of suitable housing locations. The latter is done by spatially depicting general expert preferences.

The accuracy in finding the most and least technical suitable locations is dependent on the reliability of the input data. The data used in this study originated from different sources and therefore also showed different levels of accuracy. For that reason, the resulting maps are less accurate than the least accurate layer used in their composition. As the methods have been successfully tested under German as well as Italian data conditions, it may be assumed that they can be applied in many European countries.

Since the calculation of the solar energy potential estimation depends on the application of the *r.sun* model and on the *pvgis* data, it can be applied in every region. However, the accuracy depends on input data (DEM) and on *pvgis* data availability.

Data of wind speeds are also available for every country. The only difference between the German and the Italian wind speeds was that the Italian data are calculated at a height of 25 m and the German ones at a height of 10 m, a height which better suits the requirements of microgeneration.

Wind speeds can deliver a good approximation to the wind energy potential, but for the planning of new settlements, a simulation of the wind flow would be more useful.

The accuracy of geothermal energy estimation is dependent on the data availability (e.g., the profiles) of the rock layers under the ground. This study demonstrated that even if no data on stratification and soil characteristics have been available, the information needed can be generated by the assistance of geologists with local knowledge. Nevertheless, it should be restricted to the creation of suitability maps for the use of horizontal loops. For more precision, it will be necessary to conduct further specific studies. It will also be important to have more detailed data regarding the groundwater level and movement in order to estimate the geothermal energy potential using the groundwater flow.

The biomass potential estimates can be calculated in every region. Only the data about forested areas which are available for biomass use or short rotation coppice are required. The next step will be to estimate the wood extraction capacity and transportation costs.

The best energy mix, calculated after normalization and overlay, could be derived more precisely if other criteria were included, e.g., costs, local characteristics, as well as electricity and heat needs. Such results would better support the practical planning. Methodologically, multicriteria evaluation methods in a GIS could be used for this optimization of settlement allocation with respect to sustainability.

Expert weighting of criteria about the location of an energy-efficient residential development in combination with the use of GIS and multicriteria analysis were useful for supporting the complex planning process. Various experts independently came to a considerable degree of agreement about their general preferences. The proposed method offers some advantages over the classical site suitability analysis techniques: First, it provides a structured approach to derive the suitability by 'decomposing' a complex problem into three levels (energy potentials, expert preferences for housing development, expert preferences for housing development with micro-generation technologies). This allows planners and public authorities to focus on a systematic analysis of the factors for each level. A disadvantage is that the criteria are less differentiated than in a conventional environmental impact or suitability assessment. Also, supplementing with new criteria needs considerable effort. Second, this method allows for incorporating criteria, which differ in nature. Furthermore, the method is a suitable way to weight the different criteria if no democratic legalized standards are available as a basis for weighting and decision making. Third, the approach provides an opportunity for decision makers to incorporate their own judgments. However, for a transparent practical application, the general expert preferences, which substitute the legal valuation and assessment standards, have to be presented separately from the preferences of local politicians and stakeholders. Fourth, the general preference and not a special site-specific individual interest is relevant, which may help at the same time to support rational decisions, in particular in local development and achieve a good acceptance of the results. Fifth, if regional/local stakeholder preferences are taken as a basis, the methods can be used in order to model the probable future expansion of housing development according to local interests. If mandatory zoning is weak or non-existent, land use planning can use this information for strategy building.

In the future, more concrete legal standards and priorities for decisions about energy-efficient housing and the environment may more strongly confine the importance of the expert preferences. In that case, both more predefined priorities should be included in the method and their combination with conventional impact assessment should be recommended.

Environmental planners and public authorities often make complex decisions within a short period of time when they must take into account sustainable development and participation. A set of land-use suitability maps (e.g., as part of a landscape plan) would be very useful for supporting fast decisions. Once the maps are available, land planners can analyze any new project by using simple operations such as map overlay or statistical analysis of a given area.

Clearly, the criteria selected for housing development and for micro-renewable preferences need to be combined with other siting factors which are relevant on broader scales. In this context, it will be interesting to compare the landscape plan of Sardinia and the land use plans of the municipalities with the results obtained in this work to identify the benefits and limitations. For further research, we will also integrate a few territorial and landscape-geographical aspects on a larger scale (e.g., a local scale) in order to compare them with those of the energy potentials. The aim is to give a more complete assessment to support planning decisions by integrating relevant territorial, environmental, and landscape criteria for assessing the new housing development.

CONCLUSIONS

The need to reduce the oil consumption and to produce renewable energy favors the integration of micro-renewable energy generation into housing development. Urban and regional planning can optimize this integration by selecting the best suited areas with the highest energy potential and the least environmental impacts as well as by choosing the best mix of renewables for each individual residential site. This paper suggests a methodology for finding the best locations for new housing developments which use micro-renewable technologies. The results obtained are of direct relevance for practical planning in different European countries. The methodology proposed can be an effective tool for planners in Europe to simulate new residential areas and evaluate their energy potentials for tracking changes and identifying the best solutions.

AUTHORS' INFORMATION

CP received the title of Doctor Europeus from the universities of Cagliari and Hannover in 2011. Her research interests are centered on the integration of micro-renewable technologies in regional and urban planning.

ACKNOWLEDGMENTS

This study of the integration of micro-renewable technologies in regional planning was financially supported by the Italian Ministry of Education, University and Research (Ministero dell'Istruzione, dell'Università e della Ricerca (MIUR)).

COMPETING INTERESTS

The authors declare that they have no competing interests.

AUTHORS' CONTRIBUTIONS

CP conceived and carried out the study and drafted the manuscript. She worked on all parts and in particular on the adaptation and development of methods for the estimation of the micro-energy potentials. EA helped define the general methodology of the study, participated in the design, landscape criteria selection, and planning in the eastern metropolitan area of Cagliari, and supervised the survey in Italy. AL helped develop the structure of the paper and supervised the multicriteria analysis, GIS work, and the survey in UK. CvH helped define the general methodology of the study, participated in the environmental impact selection of micro-renewable technologies, landscape, and regional planning, and supervised the survey in Germany. All authors read and approved the final manuscript.

REFERENCES

1. EU (2010) Renewable energy: *White Paper laying down a Community strategy and action plan*. http://europa.eu/legislation_summaries/other/l27023_en.htm. Accessed 21Sep 2010
2. Macleod A: **Using the microclimate to optimise renewable energy installation.** *Renewable Energy* 2008,**33**(8):1804–1813. 10.1016/j.renene.2007.
3. Zedfactory: *Projects.* 2010. http://www.zedfactory.com/about.html. Accessed 21 September 2010
4. Deleske A Stadtteil Vauban, Freiburg; 2010.http://www.vauban.de/. Accessed 21 September 2010
5. Jenks M, Dempsey N: *Future forms and design for sustainable cities*. Elsevier/Architectural Press, Oxford; 2005.
6. Droege P: *The renewable city: a comprehensive guide to an urban revolution*. Wiley, Chichester; 2007.
7. Brookes L: **Energy efficiency fallacies—a postscript.** *Energy Policy* 2004, **32:** 945–947.
8. Linden A-L, Carlsson-Kanyama A, Eriksson B: **Efficient and inefficient aspects of residential energy behaviour: what are the policy instruments for change?** *Energy Policy* 2006,**34**(14):1918–1927.

9. International Energy Agency (IEA): : *Key world energy statistics. 2007.*. http://www.iea.org/textbase/nppdf/free/2007/key_stats_2007 .pdf. Accessed 21 September 2010.

10. Maxwell EL, Renne DS: *Measures of renewable energy resources.* NREL/MP-463–6254. National Renewable Energy Laboratory, Golden, Colorado; 1994.

11. Ivanov P, St L, Trifonova L, Renne D, Ohi J: **An investigation of renewable resources and renewable technology applications in Bulgaria.** *Environ Manage* 1996,**20**(Suppl 1):583–593.

12. Schneidera DR, Neven D, Zeljko B: **Mapping the potential for decentralized energy generation based on renewable energy sources in the Republic of Croatia.** *Energy* 2006,**32**(9):1731–1744.

13. Vettorato D, Zambelli P: **Estimation of energy sustainability at local scale.** In *45th ISOCARP Congress 2009*. University of Porto, Porto; 2009.

14. Jankowski P, Nyerges T: *Geographic information systems for group decision making*. Taylor & Francis, London; 2001.

15. Malczewski J: **GIS-based multicriteria decision analysis: a survey of the literature.** *Int J Geogr Inf Sci* 2006,**20**(7):703–726.

16. Bredemeier B, Dandova M, Jähnchen I, König T, Liang X, Palmas C, Peschel I, Stegemann J: *Mikrogeneration erneuerbarer Energien. Energiepotenziale in der Region Hannover (Micro generation from renewables. Energy potential in the Region of Hannover). Master project.* Department for Environmental Planning, Faculty of Architecture, Hannover University, Hannover, Alemanya; 2009.

17. Rode M, Schneider K, Ketelhake G, Reisshauer D: *Naturschutzverträgliche Erzeugung und Nutzung von Biomasse zur Wärme- und Stromgewinnung. Ergebnisse aus dem F+E-Vorhaben 80283040 des Bundesamtes für Naturschutz*. Bundesamt für Naturschutz, Bonn; 2005.

18. Voogd H: *Multicriteria evaluation for urban and regional planning.* Pion, London; 1983.

19. Nijkamp P, Voogd H: **A survey of qualitative multiple criteria choice models.** In *Measuring the unmeasurable*. Edited by: Nijkamp P, Leitner H, Wringley N. Kluwer/Nijhoff, Dordrecht; 1986.

20. Eastman JR, Kyem PAK, Toledano J: *GIS and decision making*. UNITAR, Geneva; 1993.

21. Malczewski J: *GIS and multicriteria decision analysis*. Wiley, New York; 1999.

22. Geneletti D: **Formalising expert's opinion through multi-attribute value functions: an application in landscape ecology.** *J Environ Manage* 2005, **76:** 255–262.

23. Janssen R, Rietveld P: **Multicriteria analysis and geographical information systems: an application to agricultural land-use in the

Netherlands. In *Geographical Information System for urban and regional planning*. Edited by: Scholten HJ, Stillwell JCH. Kluwer Academic Publishers, Dordrecht; 1990.

24. Eastman JR, Jiang H, Toledano J: **Multi-criteria and multi-objective decision making for land allocation using GIS.** In *Multicriteria analysis for land-use management*. Edited by: Beinat E, Nijkamp P. Kluwer Academic Publishers, Dordrecht; 1998.
25. Banai R: **Fuzziness in geographic information systems: contributions from the analytic hierarchy process.** *Int J Geogr Inf Syst* 1993,**7**(4):315–329.
26. Carver SJ: **Integrating multicriteria evaluation with geographical information systems.** *Int J Geogr Inf Syst* 1991, **5:** 321–339.
27. Joerin F, Theriault M, Musy A: **Using GIS and outranking multicriteria analysis for landuse suitability assessment.** *Int J Geogr Inf Sci* 2001,**15**(2):153–174.
28. Pereira JMC, Duckstein L: **A multiple criteria decision-making approach to GIS based land suitability evaluation.** *Int J Geogr Inf Syst* 1993,**7**(5):407–424.
29. Jankowski P: **Integrating geographical information systems and multiple criteria decision making methods.** *Int J Geogr Inf Syst* 1995,**9**(3):251–273.
30. ISTAT: *Dati demografici (demographic data)*. 2010. http://www.istat.it. Accessed 21 September 2010
31. EU Joint Research Centre (JRC): *Solar model r.sun*. 2010. http://re.jrc. ec.europa. eu/pvgis/solres/solmod3.htm. Accessed 21 Sep 2010
32. Suri M, Hofierka J: **A new GIS-based solar radiation model and its application for photovoltaic assessments.** *Trans GIS* 2004,**8**(2):175–190.
33. Touma JS: **Dependence of the wind profile power law on stability for various locations.** *Air Pollution Control Association* 1977, **27:** 863–866.
34. Counihan J: **Adiabatic atmospheric boundary layers: a review and analysis of data from the period 1880–1972.***Atmospheric Environment* 1975, **79:** 871–905.
35. Kaltschmitt M, Huenges E: *Energie aus Erdwärme*. Deutscher Verlag für Grundstoffindustrie, Stuttgart; 1999.
36. Gerlinger E, Gerlinger E: *oral. Biokompakt*. 2008. www.biokompakt. com. Accessed 21 Sep 2010
37. Ueberhuber CW: **Monte Carlo techniques.** In *Numerical computation 2: methods, software, and analysis*. Springer, Berlin; 1997.
38. Cheney W, Kincaid D: *Numerical mathematics and computing*. 5th edition. Thomson Learning, Belmont; 2004.
39. Saaty TL: *The analytic hierarchy process*. 3rd edition. McGraw-Hill, New York; 1980.
40. *ESRI*. 2010. http://www.esri.com/ Accessed 16 Jun 2012

41. Genoa University: *CESI Research Center (2002) Italian Wind Atlas (Atlante Eolico d'Italia)*. http://atlanteeolico.rse-web.it/viewer.htm. Accessed 13 Jun 2012
42. Verein Deutscher Ingenieure (VDI): *Thermische Nutzung des Untergrundes.–Grundlagen, Genehmigungen, Umweltaspekte*. Blatt 1, Düsseldorf, Richtlinie 4640; 2010.
43. Aru A, Baldaccini P, Vacca A: *Nota illustrativa alla carta dei suoli della Sardegna (Explanatory note to the soil map of Sardinia). Regione Autonoma della Sardegna, Università degli Studi di Cagliari*. Region of Sardinia, Cagliari University; 1991.
44. Sass I, Burbaum U, Petrat L: *Staufen im Breisgau: Artesisches Grundwasser, Anhydrit und Karst im Konflikt mit geothermischen Bohrungen.* Edited by: Schwerter R. 17. Tagung für Ingenieurgeologie, Zittau; 2009.

3

GEOVISUALIZATION: A NEWER GIS TECHNOLOGY FOR IMPLEMENTATION RESEARCH IN HEALTH

Sandul Yasobant[1], Kranti Suresh Vora[1], Carl Hughes[2], Ashish Upadhyay[1], Dileep V. Mavalankar[1]

[1]Indian Institute of Public Health-Gandhinagar, Ahmedabad, India

[2]McGill University, Montreal, Canada

ABSTRACT

Use of GIS in public health is currently limited to tracking immunizations, disease surveillance, and establishing new service areas. The potential of GIS to improve planning, implementation and monitoring of programs and to inform policymaking processes for universal access to healthcare is grossly underutilized. Major reasons for restricted use are paid access to GIS software and difficult interpretation for non-GIS professionals. WebGIS technologies present an opportunity for non-GIS public health professionals to present complex data and findings in simpler manner. Conventional GIS methods and

Geovisualization are compared and contrasted in this paper using data from the MATIND project in the Gujarat state of India. In-depth literature review on GIS techniques used in health implementation research was conducted. In this paper, MATIND data have been used for comparing conventional GIS methods with a newer web based GIS tool-Geovisuali- zation. GIS software is more useful for analytical purposes; whereas web based geovisualization techniques are much better for visualization and easy interpretation of results by non GIS public health researchers. As the images are easy to interpret, interactive/dynamic and not much technical expertise is required to perform basic analysis, the tool is useful for policy makers and planners. Geovisualization provides a user-friendly tool for presenting large scale community based survey data. Increased use of this tool will help to present implementation research in a creative way to the program planners and policy makers. For program managers, it is a useful tool to monitor implementation and impact of a program to improve health of population.

Keywords: Geovisualization, GIS, Public Health

1. INTRODUCTION

Complex and multidimensional data are examined to assemble meaningful information to improve planning, implementation and monitoring of programs and evidence based policymaking processes to improve population health [1] . Various advanced methodologies are used by program managers and researchers for examining data to serve specific purpose. As public health datasets such as Health Information and Management System (HMIS) and larger health surveys such as Demographic Health Surveys have become increasingly complex; there is a growing need for innovative data analysis methods and data representation tools for optimal utilization of information [2] [3] . A number of quantitative approaches have been used to address the complex analysis of these datasets [4] [5] but critical analysis and visualization methods of spatial data are limited in public health research.

Recently, Geographical Information System (GIS) has emerged as an innovative and important component of research and practice in public health [6] . GIS has proved to be useful for various research purposes including epidemiological surveys/investigation, implementation research, program/policy decision making and dissemination of information. Despite the advancement of computational and GIS technologies [7] ; there is an urgent need for novel approaches to represent geographic data in a visual form that can improve pattern recognition/trends and hypothesis generation [8] . More importantly, it should allow for better understanding and wider usage by non-GIS experts working in public health. Visual representations can often communicate information effectively and help decision makers prioritize the actions required to improve public health outcomes [9] . Maps are an efficient means for communication, analysis, synthesis, and exploration, of geographic data and information [10] . Traditional GIS visualization techniques focus on the presentation of points, lines and polygons in static maps, such as, paper based maps. Newer techniques such as web-based geovisualization allow users to explore specific phenomena in more efficient and effective ways to uncover or clarify dimensions that may be ambiguous in traditional GIS visualizations [11] .

Geovisualization is increasingly being used to inform public health research, planning and decision making in developed countries [8] . It aids etiologic investigations and distribution of diseases, optimal deployment of limited resources and policy/regulation adoption [12] . However, the best applicability for geovisualization in public health would be for spatial analysis to get comprehensive view that is easy to understand, for complex datasets [13] . Geovisualization is often underutilized owing to a lack of training, being relatively new GIS techniques and limited landmark examples in public health. Developing countries with weak health systems and higher burden of diseases that require such tools to inform policy making and program planning to maximize limited resources have even further underutilization [14] . In India, the disease pattern is complex with current epidemiological transition and coexistence of non-communicable and communicable diseases. There is limited evidence from low resource settings on how to design simple, functional geovisualization applications for difficult policy making and program prioritization [15] . To address the gap, this paper aims to demonstrate the potential of geovisualization technique and compare it with the traditional GIS methods using data from MATIND project in the Gujarat state of India.

2. OVERVIEW OF GIS TECHNIQUES

2.1. Technique for Bringing Health Data in to GIS

Applications of GIS methods require accurate transformation of location information from the data into geographic objects; this process is known as geocoding (latitude and longitude). Geocoding which is transformation of quantitative data to GIS data is not only tough work but also very tedious and base pillar for any visualization with its rewards. Geocoded data can be processed in three major ways before visualization as described here:

2.1.1. Point Patterns

Attempt to display of the data points such as public and private health facilities, mothers' home, village settlements etc. It is useful for defining areas of case occurrences, visual inspection of spatial clusters of patients or health facilities and analyzing health care resources distribution. An example of point pattern analysis in maternal health is the geographic uptake of a government scheme such as "Chiranjeevi Yojana" (CY) in the MATIND data.

2.1.2. Line Patterns

Vectors or road lines are graphic resources that aid understanding the proximity and accessibility of patients to health care facilities. Arrows with widths proportional to the volume of flow between areas are important tools to evaluate the health care utilization of different locations.

2.1.3. Area Patterns

For spatial descriptive analysis, the administrative boundary like village, block and District data on maps, with the variable of population, no. of house hold and no. of CY users can be used to represent the data in a specific dimension.

2.2. Traditional GIS Visualization Technique

Frequently used maps in the public health research are Dot-density and Chloropleth maps [16] . Dot-density maps are the simplest way to display events. These maps use dots or other symbols to represent the number of occurrences of a given data characteristic [17] . Each dot or symbol used on the map represents a single entity (one dot = 1 case) or a group (one dot = 1000 cases). Dot-density maps are useful for area comparisons. However, dot-density maps need to be interpreted with caution for the "symbol to data characteristic" ratio. An example of a dot-density map for MATIND CY users has been shown in "Figure 1", where each dot represents a single mother or a health facility providing obstetric care.

Choropleth maps are area maps in which polygons are shaded, colored, or patterned according to the value of a given attribute for each polygon. Choropleth maps are also called thematic maps or Shaded maps. An example of a choropleth map for concentration of CY users in a district is shown in "Figure 2". The same information from Figure 1 is presented as shaded areas where darker areas represent higher concentration of CY users.

It is important to choose the right features for map presentations as the choice of color, pattern, size, polygon shape, and class intervals can impact how one interprets the information presented in a map. Single-color maps with varying color intensity (shades) are often an effective means of presenting data, but the use of differing patterns can help a black-and-white or grey-scale map. Similar-size polygons are recommended to the extent possible, as a few large polygons can dominate a map, leading to misinterpretation of information. Proportions or rates can be displayed by different class interval schemes, such as equal intervals (equal ranges of values) or quintiles (equal number of polygons falling into each class defined by dividing the range of values). The latter technique is particularly useful for presenting skewed data where distribution of utilization is not known or is very different.

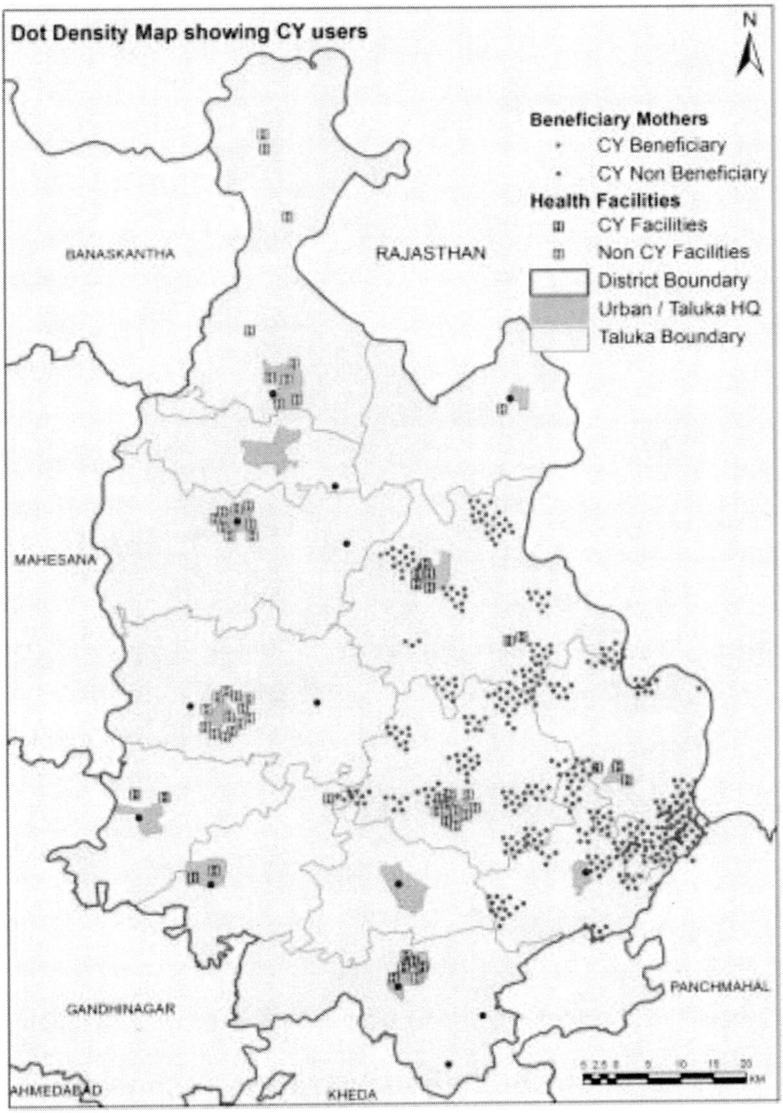

Figure 1. Dot density map showing CY users.

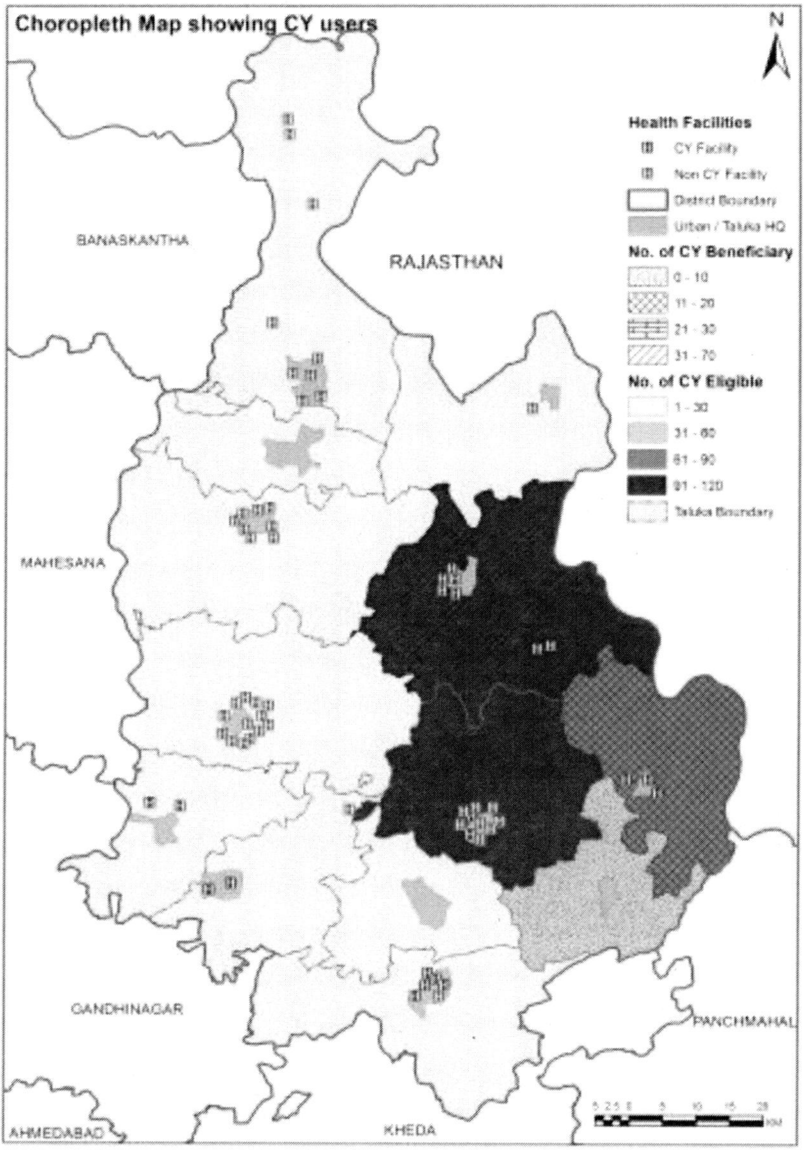

Figure 2. Choropleth map showing CY users.

2.3. Web Based Geovisualization Technique

Geovisualization is the process of geospatial data analysis where visualization is enabled through tools by the convergence of information, cartographic and geographic methods [14]. Specific use of this technique is for geospatial data displays to explore, analyze and synthesize data for generating hypotheses and developing solutions, and comprehensive representation of data. There are two philosophical approaches to geovisualization, the positivistic and phenomenological approach. The positivistic approach uses spatial modeling to best possible represent the real world. The phenomenological approach allows for individual interpretation of space and time and tends to represent space in the abstract [18] .

Addition of novel geo-web tools such as application programming interfaces (APIs) and open-source geospatial analysis coding packages in the Web 2.0 has further improved the utility through the positivistic approach [19] . Web 2.0 is distinctive with functions such as interaction, attribute filtering (including spatial and temporal attributes), dynamic and animated displays, near-real time data updates and advanced analysis. Because of the flexibility and utility of geovisualization, the only situation where it cannot be used is for non-spatial data [14] .

In the context of WebGIS, data representation is called "the science of virtual space", or the way space is constructed, reconstructed, represented and finally interpreted through a specific method [20] . The degree of realism or abstractedness in the representation of symbols as well as space must be directly connected to the intended purpose of the tool. Owing to the size and complexity of datasets, computer integration is necessary to facilitate geographic inquiry. When implementing computational methods, for example, patterned aggregation of features by attribute, the synthesis of false relationships by the user must be avoided by optimizing the integration [20] . The geovisualization interface provides a bridge between the user and knowledge construction from the data and is considered an external representation in that interacting with the tool will alter the information that is displayed subsequent knowledge that is created [20] . Finally, different user groups will have different user experiences so the target user group needs to be a central consideration for tool design. The choice of data representation, degree of computational integration and interface design can significantly alter the cognitive usability for different users and must be considered together for a comprehensive design [21] . In order to optimize its utility, the design of a geovisualization should be based on rigorous methods of data representation, visualization-computation integration, user interface and matching of the tool with the intended purpose [22] .

Geovisualization Technique Used on MATIND Data

Python program was used to extract the MATIND data from the REDCap [23] (Research Electronic Data Capture is a secure, web-based application designed to support data capture for research studies) database and to convert the data into spatial information. The program utilized the REDCap API to extract the latest version of the database through the use of a unique API URL and key. The dataset was then parsed using the simple KML python module to create point

KML features for each individual mother's home location as defined by the centroid (geographic centre) of their home village. Separate KML layers, each indicating the outcome of a certain health variable for each mother (for example, CY status, ANC check-ups, type of delivery etc.), were also created specifically for the research objective requirements. Additionally, aggregated village polygon KML layers were created of the same indicators to represent regional level outcomes. The symbology of the layers (color gradient or symbol size/icon) was chosen based on the requirement. Once the spatial layers were created, and automatically uploaded to a cloud based server (Google Drive) where they could be accessed for the geovisualization as shown in "Figure 3".

The best part of the tool is hosting large datasets on the cloud that can be accessed, edited and updated remotely and no longer needs to be on an individual hard drive or server. Subsequently, the data can be created and/or managed as well as accessed and utilized by a group of actors including: governments, citizens, civil society, businesses or academia on a variety of platforms without needing special software. This is largely due to the convergence of geospatial data standards (led by the Open GIS Consortium) promoting interoperability between and within traditional GIS methods and Web-GIS applications such as geoweb [21] . Users can create online maps on services such as the Google Maps or Open Street Maps platform with very basic knowledge of any one of a multitude of computer languages. Furthermore, cloud and non-Cloud based datasets can be dynamically linked to these services in a variety of methods and spatial data formats.

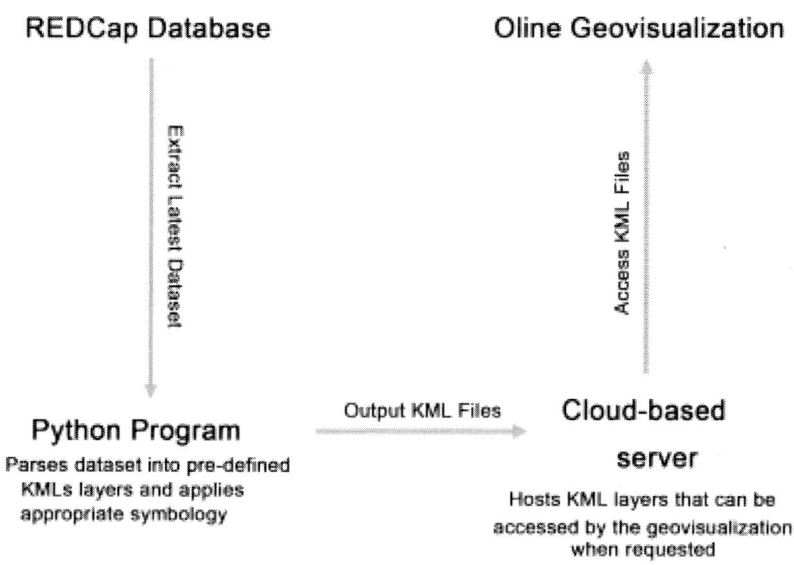

Figure 3. Process of development of web based geovisualization.

"Figure 4" shows a template of web based geovisualization. From this step on, a non-GIS expert can access and use the spatial data as per their need without any requirement for GIS software or extensive training.

Here geovisualization is an HTML file (webpage) that uses the Google Maps API platform to display spatial data. A drop down menu of individual and aggregate level spatial indicators displayed in the bottom corner on the right side of the webpage allows the user to select the indicator/layer of their choice. As users explore the data, they are able to view all the attributes of each point or polygon by clicking the point of interest and viewing its attributes in a side window. To facilitate usability and reduce cognitive issues pertaining to cluttering the user is able to zoom and pan throughout the map, as well as change the style of the map (satellite, road map, custom, etc.)

Both conventional GIS methods and Geovisualization have their advantages and limitations. Decision of which technique to be used, depends on the purpose. For example, conventional GIS techniques are better for in- depth spatial data analysis while Geovisualization has superior spatial data display.

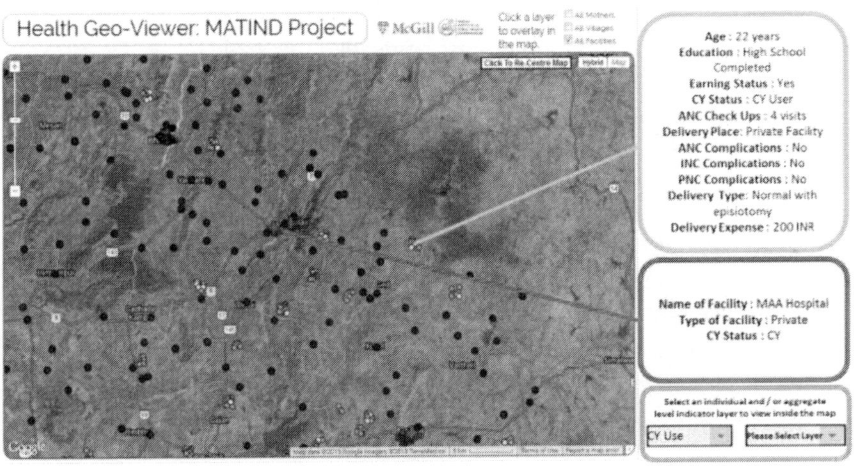

Figure 4. Web based Geovisualization snapshot of spatial distribution of CY users.

3. COMPARISON OF CONVENTIONAL GIS TECHNIQUE WITH GEOVISUALIZATION

Currently, the scope of using the technique of geovisualization is widening from exclusive use by GIS experts, innovators and early adopters towards a broader audience of non-GIS users including epidemiologists, policy makers and program planners. MATIND Gujarat data show that Geovisualization allows a number of variables to be displayed in a single view and gives the user an improved understanding of the complex relationship between these variables. For example, to understand the dynamic relationship between CY use and its predictors in the study areas such as proportion of CY eligible women, number of CY providers and their spatial distribution (as seen in Figure 3). This in turn can be useful to identify gaps and ways to improve implementation of program in a local context. "Table 1" compares newer geovisualization with traditional GIS soft-wares from different aspects.

Future of Geovisualization in Implementation Research in Health

As Slocum et al. suggest, "The most sophisticated technology will be of little use if people cannot utilize it ef- fectively" [17] . In the same vein, widespread use of geovisualization by public health researchers will make the Table 1. Comparison of traditional software based GIS techniques with geovisualization.

Complex and labor intensive process of development of the tool worthwhile. This could be extended to the different areas of public health to evaluate the unmet needs by geographic areas which are important for realistic evaluation and to prioritize resource allocation for low resource environments such as India for optimal use and maximum possible coverage. In single line the conclusion is that GIS technologies work hard for program managers and planners without hard work by program managers and planners to learn advanced GIS techniques. Apart from ease of use, it is cheaper option to software based GIS technology and once created can be used for a long time by multiple stakeholders.

4. CONCLUSION

Public health experts are using GIS techniques sparsely due to difficulty in learning the technology and the cost of software. Geovisualization provides a user-friendly tool for presenting large scale community based survey data or routinely collected HMIS data without losing the complexity. Dynamic, interactive, and temporal geovisualization make it possible for non-GIS experts to understand and disseminate public health data which are inherently spatial in nature and cannot, as easily, be presented through the use of paper based GIS maps or by quantitative analysis only.

AUTHOR CONTRIBUTION

Conceptualization: SY, KV, DVM. Analysis: SY, CH, AU. Writing: SY, KV, CH. Review and critical comments: KV, DVM.

CONFLICT OF INTEREST

The authors declare that they have no competing interests.

SOURCE OF FUNDING

Sandul Yasobant, Kranti Suresh Vora, Carl Hughes, Ashish Upadhyay, Dileep V. Mavalankar (261304,30)">261304].

REFERENCES

1. Andrienko, G., Andrienko, N., Dykes, J., Fabrikant, S.I. and Wachowicz, M. (2008) Geovisualization of Dynamics, Movement and Change: Key Issues and Developing Approaches in Visualization Research. Information Visualization- Basingstoke, 7, 173-180.

2. Ashish, J. and Chiehwen, E.H. (2012) A Human Centered Geovisualization Framework to Facilitate Visual Explora- tion of Telehealth Data: A Case Study. Technology and Healthcare, 20, 473-488.

3. Bhowmick, T., Griffin, A.L., Maceachren, A.M., Kluhsman, B.C. and Lengerich, E.J. (2008) Informing Geospatial Toolset Design: Understanding the Process of Cancer Data Exploration and Analysis. Health & Place, 14, 576-607.

4. Bhowmick, T., Robinson, A.C., Gruver, A., Maceachren, A.M. and Lengerich, E.J. (2008) Distributed Usability Evalu- ation of the Pennsylvania Cancer Atlas. International Journal of Health Geographics, 7, 36.

5. Cinnamon, J., Rinner, C., Cusimano, M.D., Marshall, S., Bekele, T., Hernandez, T. and Chipman, M.L. (2009) Eva- luating Web-Based Static, Animated and Interactive Maps for Injury Prevention. Geospatial Health, 4, 3-16.

6. Clarke, K.C., Mclafferty, S.L. and Tempalski, B.J. (1996) On Epidemiology and Geographic Information Systems: A Review and Discussion of Future Directions. Emerging Infectious Diseases Journal, 2, 85-92.

7. Dennis, A.R. and Carte, T.A. (1998) Using Geographical Information Systems for Decision Making: Extending Cogni- tive Fit Theory to Map-Based Presentations. Information Systems Research, 9, 194-203.

8. Dykes, J., Maceachren, A. and Kraak, M. (2005) Modelling Virtual Environments for Geovisualization: A Focus on Representation. Exploring Geovisualization, Chapter 19, 389-402.

9. Elwood, S. (2008) Geographic Information Science: New Geovisualization Technologies—Emerging Questions and Linkages with GIScience Research. Progress in Human Geography, 33, 256-263.

10. Lawson, A., Biggeri, A., Böhning, D., Lesaffre, E., Viel, J.F. and Bertollini, R. (2001) Disease Mapping and Risk As- sessment for Public Health. International Journal of Epidemiology, 30, 405-406.

11. Maceachren, A.M. and Brewer, I., (2004) Developing a Conceptual Framework for Visually Enabled Geocollaboration. International Journal of Geographic Information Science, 18, 1-34.

12. Maceachren, A.M., Gahegan, M., Pike, W., Brewer, I., Cai, G., Lengerich, E. and Hardistry, F. (2004) Geovisualization for Knowledge Construction and Decision Support. IEEE Computer Graphics and Applications, 24, 13-17.

13. Maceachren, A.M. and Kraak, M.J. (2001) Research Challenges in Geovisualization. Cartography and Geographic In- formation Science, 28, 3-12.

14. Malczewski, J. (2006) GIS-Based Multi Criteria Decision Analysis: A Survey of the Literature. International Journal of Geographical Information Science, 20, 703-726.

15. Menno-Jan, K. (2003) Geovisualization Illustrated. ISPRS Journal of Photogrammetry & Remote Sensing, 57, 390- 399.

16. Robinson, A.C., Chen, J., Lengerich, E.J., Meyer, H.G. and Maceachren, A.M. (2005) Combining Usability Techni- ques to Design Geovisualization Tools for Epidemiology. Cartography and Geographic Information Science, 32, 243- 255.

17. Robinson, A.C., Maceachren, A.M. and Roth, R.E. (2011) Designing a Web-Based Learning Portal for Geographic Visualization and Analysis in Public Health. Health Informatics Journal, 17, 191-208.

18. Rogers, M.Y. (1999) Getting Started with Geographic Information Systems (GIS): A Local Health Department Pers- pective. Journal of Public Health Management and Practice, 5, 22-33.

19. Sarat, M.K. (2009) Interactive Visual Summarization of Multidimensional Data. IEEE Conference on Information Vi- sualization, San Antonio, 11-14 October 2009, 65-72.

20. Shi, Y., Song, Y. and Zhang, A. (2003) A Shrinking-Based Approach for Multi-Dimensional Data Analysis. Proceed- ings of 29th International Conference on Very Large Data Bases, Berlin, 9-12 September 2003, 1-12.

21. Van Elzakker, C.P.J.M. (2003) The Use of Maps in the Exploration of Geographic Data. Proceedings of the 21st In- ternational Cartographic Conference: Cartographic Renaissance, Durban, 10-16 August 2003, 1945-1955.

22. Thrall, S.E. (1999) Geographic Information System (GIS) Hardware and Software. Journal of Public Health Manage- ment and Practice, 5, 82-90.

23. Harris, P.A., Taylor, R., Thielke, R., Payne, J., Gonzalez, N. and Conde, J.G. (2009) Research Electronic Data Capture (REDCap)—A Metadata-Driven Methodology and Workflow Process for Providing Translational Research Informatics Support. Journal of Biomedical Informatics, 42, 377-381.

4

CONSTRUCTIONS OF RESILIENCE: ETHNORACIAL DIVERSITY, INEQUALITY, AND POST-KATRINA RECOVERY, THE CASE OF NEW ORLEANS

Kevin Fox Gotham [1,*] and Richard Campanella [2]

[1] Department of Sociology, Tulane University, 220 Newcomb Hall, New Orleans, LA 70118, USA

[2] School of Architecture, Tulane University, New Orleans, LA 70118, USA

ABSTRACT

In this paper, we draw on multi-level census data, in-depth interviews, ethnographic and Geographical Information Systems (GIS) methods to examine the effects of median household income, ethnoracial diversity, and flood damage on rates of post-Katrina repopulation in New Orleans. Our main finding is that New Orleans neighborhoods have been experiencing modest increases in ethnoracial diversity as well as a retrenchment of socio-spatial inequalities, as measured by low diversity scores, low median household income levels, and high poverty rates. In addition to documenting the objective indicators of "recovery", we draw attention to the socially constructed nature of resilience. Based on interviews and ethnographic field observations, we investigate how resident constructions of resilience shape their views of the post-Katrina recovery process, provide a compelling and reassuring story of community revitalization, and convey a sense of collective power and control despite continued vulnerability to hazards and disasters.

Keywords: resilience; disaster; ethnoracial diversity; Hurricane Katrina

1. INTRODUCTION

In recent years, urban scholars across the social and environmental sciences have focused much theoretical and empirical attention on understanding and explaining the connections between resilience and post-disaster recovery and rebuilding [1,2,3,4,5]. Probably the most comprehensive development of the notion of resilience, as expressed in studies of ecology and social systems, defines the concept as a return to normalcy or a pre-disaster state of stability or equilibrium. This "engineering" concept of resilience concentrates on stability at a presumed steady-state, and stresses resistance to a disturbance and the speed of return to the equilibrium point. Other non-equilibrium approaches view resilience as akin to adaptive capacity and explore the ability of cities to rebound, regenerate, and renew when affected by a major disaster or other perturbation [6,7,8,9,10]. Such an approach eschews a notion of resilience as reaching an end-point or finale [11,12,13,14]. In the biological, ecological, and social sciences resilience is connected to ideas of functional interdependence, innovation, and transformability [15,16,17,18,19]. The resilience approach emphasizes non-linear dynamics, thresholds, uncertainty, surprise when a system is faced with exogenous shocks. Such an approach stresses how periods of gradual change interplay with periods of volatile and rapid transformation and how macro and micro dynamics interact across temporal and spatial scales. The growing importance and pervasiveness of the resilience concept across fields—sociology, economics, ecology, environmental science, and so on—attests to the increasing popularity of the term to understand ecosystem dynamics, recovery from disaster, and social-ecological transformations [20].

Informed by theoretical and empirical studies of social-ecological resilience, and based on a multi-method analysis of several data sources, we examine the ways in which post-disaster repopulation outcomes connect with social inequalities and ethnoracial diversity, using a case study of recovery outcomes in post-Katrina New Orleans. New Orleans is important to study because it represents a city that straddles the challenges of recent exogenous shocks occurring over long-term chronic distress. Since 2005, the region has experienced three major shocks including Hurricane Katrina, the Great Recession, and the Deepwater Horizon oil spill. These traumatic events have taken place in a context of slowly evolving "slow burn" stresses such as climate warming, sea level rise, and subsidence. Moreover, New Orleans has long been known as a city with troublesome social inequalities, high rates of poverty, and omnipresent community struggles linked to class conflict and racial injustice. These endemic social problems combined with an international reputation as a place of rich culture and celebrated heritage intimate New Orleans as a place of irony, paradox, and contradiction.

Our goal is to investigate the influence of ethnoracial diversity and social inequality on post-disaster recovery outcomes in the city. Three questions motivate this study: Do traumatic events alter processes and patterns of ethnoracial diversity? How do levels of ethnoracial diversity and social inequality interact to influence post-disaster recovery outcomes? How do post-disaster recovery outcomes shape resident perceptions of resilience? We begin by situating our research questions in the context of a burgeoning interdisciplinary debate on the drives and processes of post-disaster social-ecological recovery. Next, we discuss our methods, data sources, and analytical approach. Specifically, we use GIS methods, quantitative analysis, ethnographic field observations, and semi-structured interviews to investigate the recovery process in New Orleans. The rationale of our multi-method approach is underpinned by the principle of triangulation, which implies that researchers should seek to ensure that they are not over-reliant on a single research method and should instead employ more than one measurement procedure when investigating a research problem. That is, to enhance confidence in findings we use multiple methods. We also use triangulation as a means to discover new knowledge, new processes, and new trends vis-à-vis post-disaster recovery outcomes. We then present our findings based on our multi-method research design. Finally, we discuss the implications of our research for scholarship on post-disaster recovery and rebuilding.

2. DIVERSITY, RESILIENCE, AND POST-DISASTER RECOVERY

In the fields of urban studies and urban ecology, studies of post-disaster recovery have increasingly focused on "adaptive capacity", "self-organization", "diversity", "transformability" and "resilience" as major factors that explain variation in the pace and trajectory of rebuilding [21,22]. In this conception, recovery rates vary according to the degree to which the system can build and increase the capacityfor learning, self-reorganization, renewal, and innovation when faced with disturbances [23,24]. In particular, scholars in a variety of different field have used the concept of resilience to illuminate regional change and link different types of regional stresses to patterns and processes of urban recovery [25]. For post-disaster human systems, resilience lies in the ability of groups, organizations, and institutions to access and effectively use political, economic, cultural, and natural resources for recovery, transformation, and innovation. As noted, the ability of a municipality and surrounding cities to weather a disaster, according to Wallace and Wallace, "depends on the size of social networks in its neighborhoods and on the interconnection between the social networks" ([26], p. 1). In this network conception of social resilience, "networks composed of a diversity of 'bridging' links to a diverse web of resources and 'bonding' links that build trust strengthen a community's ability to adapt to change". But networks composed only of "bonding" links can impose constraining social norms, foster group homophily, reinforce distrust of non-members and thereby erode resilience [27]. Social and institutional networks exhibit varying degrees of organization, cohesion, flexibility, and resource flows. In resilient systems, networks are interdependent, heterogeneous, collaborative, and functionally redundant, with reserve capacity achieved through duplication, interchangeability, and cross-scale interconnections [28,29].

The view of cities as complex adaptive systems animates resilience scholarship that trumpets the value of "diversity" as a source of urban resilience to urban trauma. Researchers claim that diversity is essential in buffering the impact of exogenous shocks and disturbances in cities through practices that promote institutional differentiation, functional interdependence, and land-use and land-cover heterogeneity [30]. Uneven spatial distribution of ethnic and racial groups across neighborhoods often coincides with uneven access to institutional and economic resources as well as exposure to hazards and other social problems (e.g., crime, pollution, and so on) [31,32,33,34,35]. Adger *et al.* [23] argue that urban social diversity can confer resilience to the extent that governance and management frameworks can spread risk by diversifying patterns of resource use and by encouraging alternative activities and lifestyles. Such practices are analogous to the way management of a financial portfolio is able to sustain the growth in financial markets. In addition, diversity in the form of heterogeneous policies, socio-legal regulations, and land-use patterns and practices represent a mix of assets that can build functional redundancy to enhance the capacity of systems to face economic and environmental shocks and build resilience [31]. Likewise, scholars acknowledge that practices that homogenize key elements of a system, eliminate redundancies, or concentrate resources can degrade adaptive capacities and proliferate vulnerabilities to environmental and socio-economic crises [32].

Scholars have documented how various types and degrees of diversity can influence the stability and functioning of social-ecological systems. Wallace and Wallace [26] have examined how the decline of ethnoracial diversity in New York City during the 1970s contributed to a loss of resilience and a proliferation of social problems including housing abandonment, drug use, violent crime, tuberculosis, and low-weight births. Increases in racial segregation eroded longstanding, intergenerational networks that previously linked diverse ethnic groups to strong social institutions embedded in particular neighborhoods. For example, the Italian presence in East Harlem almost disappeared, and East Harlem became Spanish Harlem. Central Harlem had health areas (aggregations of approximately 10 census tracts) that had approximately 98 percent African-American population by 1980. Increasing segregation and declining diversity meant that there were very few opportunities to support the maintenance of linkages between social networks of differing characteristics. The network-eroding effects of segregation were exacerbated by population displacement. Decades of building social, economic, and political institutions were undermined in New York and other cities through highway displacement, urban renewal relocations, and outmigration of residents to the expanding suburbs [23]. In short, declines in ethnoracial diversity translated into declines in the number and strength of networked connections and an increase in the vulnerability of groups to economic downturns, fire and disease epidemics, and fiscal crisis of public institutions.

In this paper, we do not equate post-disaster "repopulation" or "recovery" with "resilience" *per se*. A post-disaster city can experience a high rate of repopulation and reinvestment with extreme class and racial inequalities as measured by high levels of poverty, unemployment, and segregation. Figure 1 lays out our conceptual and analytical framework. Accordingly, our approach does not employ an operational conception of resilience and ask resilience "of what to what" [34]. Rather, we analyze the post-Katrina recovery process by addressing the question of how repopulation outcomes and processes influence resident constructions of resilience. This approach addresses the ways in which resilience is not just an objective category but a malleable and socially constructed term that is fashioned through peoples' efforts to promote community recovery. Understanding the ways in which post-disaster recovery outcomes interlock with social constructions of resilience is an important task given the increasing vulnerability of urban populations to hazards, disturbances, and disasters [35,36].

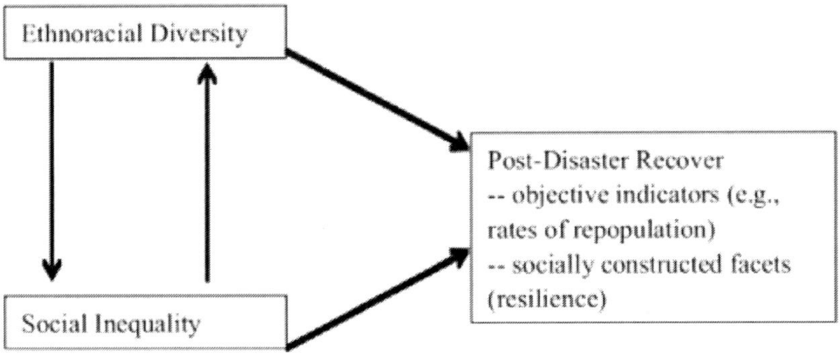

Figure 1. Conceptual and Analytical Framework.

3. METHODS, DATA SOURCES, AND ANALYTICAL APPROACH

The failure of the Army Corps of Engineers' levee system during Hurricane Katrina sparked the largest population displacement in U.S. history and radically altered the organization and character of New Orleans's urban neighborhoods. The hurricane caused catastrophic property damage along the Mississippi and Alabama coasts with 90,000 square miles of the Gulf Coast region designated as federal disaster areas, an area almost as large as the United Kingdom. The levee breaches flooded over 80 percent of the urbanized footprint of the city of New Orleans on the east bank of the Mississippi River, with some areas under twelve feet of water into September 2005. Additionally, some of neighboring Jefferson Parish to the west and all of St. Bernard Parish to the east suffered deep deluges. Nearly 228,000 occupied housing units, reflecting more than 45 percent of the metropolitan total also flooded. This total included 120,000 owner-occupied units (39 percent of the metropolitan area total) and 108,000 units occupied by renters (56 percent of the metropolitan total). More than 70,000 elderly people and 124,126 children—57,300 of them residing in single-parent families—lived in the flooded areas ([37], pp. 14–15). In short, the intensity of Hurricane Katrina's destruction, unequal effects, and disproportionate impacts on the disadvantaged makes a study of post-Katrina New Orleans timely and important.

The map (Figure 2) shows Katrina-induced flooding by its depth as measured by the Federal Emergency Management Administration (FEMA) on 2 September 2005, four days after the hurricane and levee breaches, and two days after the deluge peaked within the bowl-shaped city. The map shows flooding as measured at one foot increments overlaid with U.S. Census blockgroups ($N = 466$)—the aggregating unit by which we conduct our analysis—and parish boundaries courtesy of Environmental Systems Research, Inc. (ESRI). The map shows that the darker the area, the deeper the flood water. Light gray areas reflect lower levels of flooding. We use flood depth as a measure of flood damage. FEMA studies, which assess structural damage produced by flooding, have noted that flood water that reaches 24 inches can carry away most automobiles and cause costly damage to property [38]. Consistent with past research [39,40], we defined flood damage in terms of water depth and constructed an ordinal variable of "no flooding" (under six inches on 2 September 2005); "moderate flooding" (0.5 inches to 2.9 feet) and "deep flooding" (over three feet).

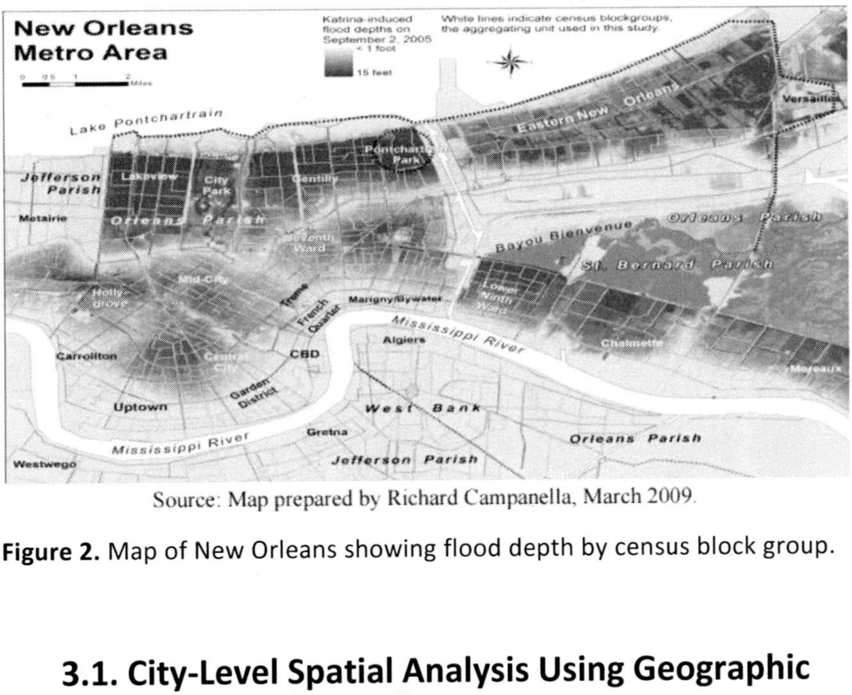

Source: Map prepared by Richard Campanella, March 2009.

Figure 2. Map of New Orleans showing flood depth by census block group.

3.1. City-Level Spatial Analysis Using Geographic Information Systems (GIS)

Our principle neighborhood-level variable of social diversity—ethnoracial diversity—is based on racial and ethnic composition. We use the Gibbs-Martin [41] index to measure social diversity, defined as:

$$D = 1 - \sum_{i=1}^{N} P_i^2$$

where *p* is the proportion of individuals in the nominal racial/ethnic category in a given census blockgroup. We use this diversity index as a statistic to measure the probability that two randomly selected individuals from a given geographical area will be of different racial/ethnic backgrounds. If all people of a given area are of the same racial/ethnic background, the diversity index is zero (perfect homogeneity). Indices closer to 1 indicate a highly diverse racial/ethnic composition, toward perfect heterogeneity. We use the index as a measure of racial mixing and concentration, e.g., low levels can be seen as evidence of racial clustering and lack of interaction and contact with other ethnic and racial groups. High levels indicate a variety of ethnoracial groups and ethnoracial networks and connections.

Data on demographics before and after Hurricane Katrina are derived from the 2000 and 2010 U.S. Census of Orleans Parish aggregated at the block level, but analyzed at the blockgroup level because it was at that spatial scale that median household income (MHHI) was posted. Because of numerous micro-changes in blocks during this ten year period, we decided to use the 2000 blockgroup polygons as the spatial analysis unit. Scholars have noted that smaller units of analysis (blockgroups) tend to be modestly more homogeneous than larger ones (census tracts) [42]. A major advantage of blockgroup analysis includes the ability to measure diversity in sub-neighborhood areas and to achieve a maximum degree of geographic drill-down enabling the most micro-perspective of demographics for a neighborhood or part of a study area. Our nominal categories include non-hispanic white, black, Hispanic, Asian, American Indian, Hawaiian/Pacific Islander, Multi-race, and Other. The vast majority of the New Orleans population fell into the first four categories (see Table 1).

Table 1. Total population and ethnoracial population in New Orleans, 2000–2011.

Orleans Parish				
	2000		**2010**	
Total Population	484,674		360,740	
Ethnoracial Diversity				
White alone, not Hispanic	128,871	26.6%	112,347	31.1%
Black or African American alone, not Hispanic	323,392	66.7%	213,489	59.2%
Hispanic (Any Race)	14,826	3.1%	18,927	5.2%
Asian alone, not Hispanic	11,007	2.3%	10,583	2.9%
Other, not Hispanic	6,666	1.4%	5,394	1.5%

Source: Elaine Ortiz and Allison Plyer. 20 September 2012. *Who Lives in New Orleans and the Metro Area Now?* Greater New Orleans Community Data Center (GNOCDC).

We back-aggregated the 2010 block-level demographics into the 2000 blockgroup polygons, using a "spatial join" (overlay) function in ESRI ArcMap 10.0 Geographical Information Systems (GIS). This process united the tabular attributes of the 2010 data with their spatial counterparts in the 2000 data, such that the output file had the spatial extents of the blockgroups as they stood in 2000 but the demographics of that area in both 2000 and 2010. The vast majority of blockgroups were identical in both decennial years; however, their internal reference numbers had been changed, which disallowed a traditional "database join" using a key item and necessitated a spatial join. All results were carefully checked for successful transmission, and no errors were found.

A detailed measurement of Katrina flood depth conducted by FEMA on 2 September 2005 was handled in a similar fashion, for example, by back-aggregating the vectorized postings of flood depth and averaging them within each Year-2000 blockgroup. This process provided a sound metric of level of damage, since the vast majority of Katrina destruction came from flood rather than wind, and the deeper the water, the greater the damage. The Gibbs-Martin index formula was then computed on the temporal pair, which was then differenced by subtracting the older metric from the more recent. The greater the magnitude of the positive number, the more that blockgroup diversified; the more negative, the more homogenized it became.

We aggregated MHHI (which reflects household incomes from 1999 reported in 2000) into five ordinal categories: "well below city average" was set at less than $20,350 (which represents 75 percent of the city average of $27,133); "below city average" (75–92 percent of city average, or $20,350 to $24,962); "city average" (92–108 percent of city average, or $24,962 to $29,304); "above city average" (108–127 percent of city average, or $29,304 to $34,459) and "well above city average" (greater than 127 percent of city average, or > $34,459). The metrics from the hundreds of blockgroups analyzed were then aggregated into unique spatial categories controlling for the three flood-damage slices and the five MHHI slices, a total of fifteen categories.

3.2. Neighborhood Level Analysis Using Ethnographic Field Observations and Interviews

In addition to our GIS analysis of census bureau data, we draw on ethnographic field observations and interviews conducted with neighborhood stakeholders in the Lower Ninth Ward neighborhood in New Orleans. This qualitative data provides a supportive and complementary role to our quantitative data collection and statistical analysis. One purpose of the interview and ethnographic data collection was to uncover how associations observed in the quantitative analyses manifest themselves in residents' daily lives. A second purpose was to provide additional, in-depth information to triangulate methods and data sources to enhance validity and reliability. A third reason was to grasp and understand local residents' experiences with post-disaster redevelopment, individually and collectively, and identify the mechanisms that influence different rates of neighborhood repopulation. A final reason was to explore conceptions of resilience in resident voices to highlight tensions between idealized conceptions and complicated realities of the post-Katrina rebuilding process.

We used a snowball sampling strategy to solicit participation from 25 community stakeholders. We contacted leaders of neighborhood coalitions, civic rights groups, and local non-profit organizations. We undertook interviews to compare and contrast experiences, understandings, and framing strategies that residents use to interpret the redevelopment of New Orleans neighborhoods in the years since Hurricane Katrina. We asked questions about the history and organization of the particular urban neighborhoods, the pace and trajectory of neighborhood recovery, and the challenges and opportunities that different neighborhoods and non-profit organizations faced in the years following the levee breaks. We also asked different residents, non-profit and neighborhood leaders about their views of Katrina's impact on neighborhoods, the role of governments in responding to the disaster, and the ways in which different policies have both helped and hindered the rebuilding effort. To facilitate analysis, the transcribed interviews were coded in themes such as "perceptions of neighborhood resilience", "perceptions of the impact of government policies and programs", and "perceptions of resilience at the city level", perceptions of resilience pertaining to different social-ecological systems and institutions—e.g., health care, education, economy, political system, flood and storm control, disaster management, ecosystem, *etc.*" Finally, we used *Atlas.ti*, a qualitative data software program, to assist with coding and structuring the interview data.

Overall, our ethnographic and interview methods complement our quantitative research in a number of ways. First, ethnography and interviews offer an opportunity to compare and contrast generalizing and particularizing analyses and findings, thus offering an opportunity for triangulating data sources and methods. Identifying similar results with different types of data enhances confidence in these results. Second, the inductive nature of ethnography and qualitative research can highlight the grassroots institutions and localized interactions around which organizing for community recovery and conflict may occur, such as disaster aid for housing recovery, reinvestment in schools, policing and security/safety, and environmental sustainability. Finally, ethnography can elucidate the mechanisms and explanations not observable in census data or GIS analyses. For example, GIS analyses can highlight changes in ethnoracial diversity and reveal demographic trends. Ethnographies can show how and why these socio-spatial changes and demographic trends occur. In short, the advantage of combining quantitative and qualitative data is that it provides a way of examining, in detail, the post-disaster recovery and transformation of a city, neighborhood, or series of neighborhoods that could be difficult to infer from one data source alone.

4. ETHNORACIAL DIVERSITY AND SOCIO-SPATIAL INEQUALITY

A major question we address is whether ethnoracial diversity had a significant impact on post-Katrina repopulation rates after controlling for income and flood depth. These relationships are probed in Figure 3 and Figure 4. Each figure shows a scatterplot of the level of ethnoracial diversity on the y-axis and repopulation rates on the x-axis. Figure 3 reports on the relationship between ethnoracial diversity in 2000 and post-Katrina repopulation, while Figure 4 reports on the relationship between ethnoracial diversity in 2010 and post-Katrina repopulation. Each figure shows the average regression line and also the R^2 (explained variance) for the regression, summarizing how strong the relationship is between ethnoracial diversity and repopulation.

Figure 3 suggests that pre-Katrina ethnoracial diversity exhibited a fairly strong positive relationship ($R^2 = 0.79$) with population recovery. That is, areas that were nearly totally homogenous according to the 2000 Census, with Gibbs-Martin indices in the 0.10–0.25 range, had the weakest population recovery rates. Areas that were more diverse (with Gibbs-Martin indices in the 0.25 to 0.46 range) fared better, having lost 15 percent to 40 percent of their populations or, in the most diverse areas, having mostly maintained their populations. No area had a diversity metric greater than 0.46. This figure suggests that while some areas were more diverse than others and they fared better in recovering their populations, no area was truly heterogeneous.

When we shift our question to inspect the relationship between post-Katrina ethnoracial diversity and repopulation, the relationship intensifies to $R^2 = 0.85$ (see Figure 4). Areas that were the most diverse, relative to the rest of the city, when enumerated in 2010, also had the highest population recovery rates. The least diverse had the lowest return rates. Overall, approximately 61 percent of all 485 blockgroups witnessed increases in ethnoracial heterogeneity while 38 percent ($N = 185$) of all blockgroups saw their diversity index move toward greater homogeneity. Block groups that did not flood saw their diversity score increase from 0.386 to 0.414 from 2000 to 2010 while blockgroups that flooded deeply witnessed an increase from 0.218 to 0.224 during the same time. Overall, Figure 3 and Figure 4 reveal strong associations between ethnoracial diversity and repopulation.

In both figures the regression lines have a clearly positive slope and the explained variance is substantial.

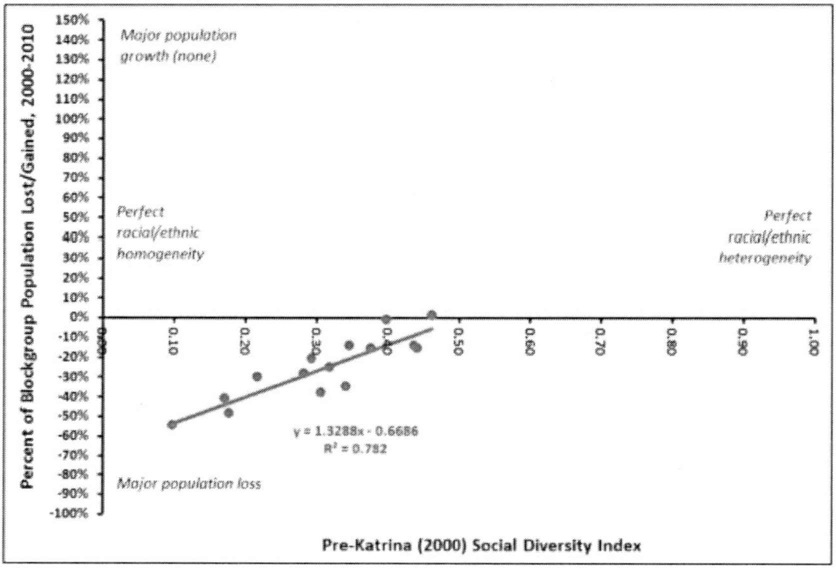

Figure 3. Relationships Between Pre-Katrina (2000) Social Diversity and Post-Katrina (2000–2010) Population Recovery.

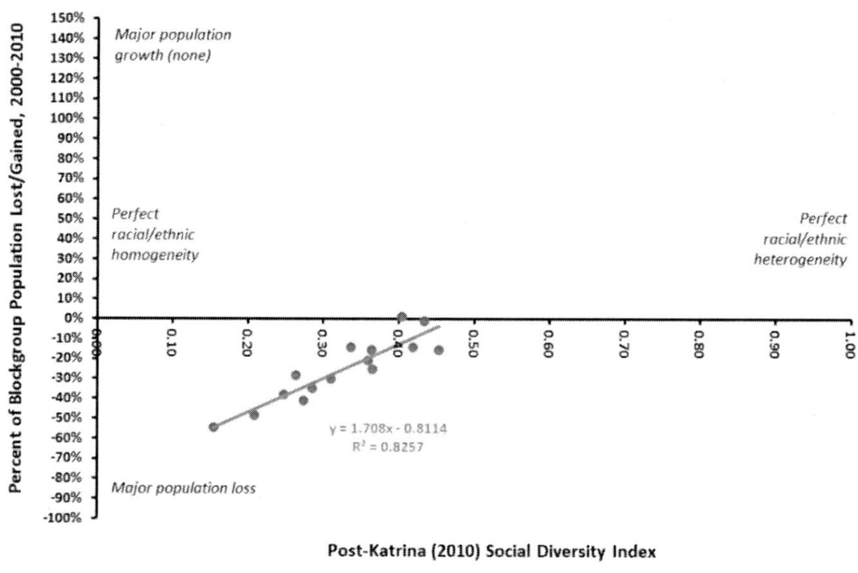

Figure 4. Relationships Between Post-Katrina (2010) Social Diversity and Post-Katrina (2000–2010) Population Recovery.

Increases in ethnoracial diversity reflect post-Katrina declines in the African American population and a growth in the number and percentage of Hispanic and Asian residents. Before Hurricane Katrina, the vast majority of the most racially homogenous areas in New Orleans were African-American, and because most African-Americans in New Orleans are significantly poorer than whites, many members of this population found themselves in a socially vulnerable position when confronted with the disaster-induced displacement [43,44]. In our analysis of blockgroup data, African American residents made up 76 percent of the population that lived in deeply flooded areas, respectively. By contrast, whites made up just 20.6 percent of those who lived in neighborhoods that flooded deeply. Additionally, African Americans also found themselves more geographically exposed to the hazard, because they predominated in the eastern half of the metropolis, which suffered more than half of the total flooding due to its canal-scared landscape and proximity to the storm surge. African-Americans were also more likely to live on lower topographic zones and areas with low base-flood elevation levels relative to whites and people of higher socioeconomic status [45]. Moreover, the return of African Americans has been sluggish because many tended to live in areas that experienced greater flooding and hence suffered more severe housing damage.

Finally, Table 2 summarizes changes in ethnoracial diversity at the blockgroup level based on flood depth and median household income. When examining the diversity index in 2000, we see that blockgroups that flooded deeply reported low ethnoracial diversity scores within each of the five income groups. When examining changes in ethnoracial diversity pre- and post-Katrina, we see that all three damage categories—e.g., none/minimal, moderate flooding, and deep flooding—in the lowest income group (well below city average) saw their ethnoracial diversity score increase though moderate and deeply flooded areas experienced higher rates of diversification than blockgroups that experienced no flooding. Overall, within each of the five income categories, blockgroups that experienced high levels of trauma (flooded deeply) were more ethnic and racially homogeneous before Hurricane Katrina than areas that experienced moderate or no flooding. In addition, areas that were more ethnic and racially diverse before the disaster tended to experience less damage.

It is important to note that these ethnoracial diversity scores are very low and increases and decreases have been quite small. Two generalizations can be made from these data. First, low levels on the ethnoracial diversity index indicate extreme social homogeneity and therefore a low probability that randomly selected individuals from blockgroups will interact with different ethnic and racial groups. Low diversity neighborhoods characterize New Orleans and reflect ethnic and racial isolation and separation rather than intermingling and inclusion. Second, low diversity scores suggest entrenched socio-spatial inequality, a pattern that reveals the centrality of racial/ethnic segregation as a major characteristic of New Orleans neighborhoods. In 2000, New Orleans ranked as one of the nation's most segregated metropolitan areas with a dissimilarity index of 0.690, a figure that remained virtually unchanged over two decades, from 0.683 in 1990 and 0.700 in 1980. By 2010, this level of segregation had declined to 0.633, a substantial decline, but a high figure nevertheless. The dissimilarity index measures the degree to which two groups are evenly spread among census tracts in a given city. Evenness is defined with respect to the racial composition of the city as a whole. With values ranging from 0 to 1, the index gives the percentage of one group who would have to move to achieve an even residential pattern—one where every tract replicates the group composition of the city. A value of 0.60 or above is considered very high. For example, a score of 0.60 for black-white segregation means that 60 percent of either group would have to move to a different tract for the two groups to become equally distributed [46].

Table 2. Median household income, flood damage, and change in social diversity index by census blockgroup, 2000–2010.

Median Household Income (MHHI)	Flood Depth	Mean social diversity index, 2000	Standard Deviation	Mean social diversity index, 2010	Standard Deviation	Change in Diversity Score	Percent Change in Diversity Score
Well below city average	None/minimal	0.29	0.18	0.36	0.21	0.07	23%
	Moderate	0.17	0.18	0.27	0.20	0.10	60%
	Deep	0.10	0.12	0.16	0.15	0.06	59%
Below city average	None/minimal	0.44	0.18	0.42	0.19	−0.02	−4%
	Moderate	0.22	0.17	0.31	0.19	0.09	43%
	Deep	0.18	0.20	0.21	0.21	0.03	18%
Average	None/minimal	0.44	0.13	0.45	0.13	0.01	3%
	Moderate	0.32	0.19	0.37	0.21	0.05	14%
	Deep	0.34	0.24	0.29	0.23	−0.06	−16%
Above city average	None/minimal	0.46	0.15	0.40	0.12	−0.06	−13%
	Moderate	0.38	0.16	0.36	0.17	−0.01	−3%
	Deep	0.31	0.23	0.25	0.21	−0.06	−19%
Well above city average	None/minimal	0.40	0.14	0.43	0.15	0.04	9%
	Moderate	0.35	0.15	0.34	0.16	−0.01	−3%
	Deep	0.28	0.16	0.26	0.15	−0.02	−6%

Before Katrina, New Orleans had the second-highest rate of African American concentrated poverty in the nation, with 37 percent of the city's African American population living in neighborhoods of concentrated poverty. While this poverty rate declined to 32 percent in 2010, poverty figures reveal striking racial differences in poverty. Poverty rates for African Americans are more than double that of whites (14 percent in 2010) and, even more striking, the poverty rate for African American children below age 17 stands at 48 percent, almost seven times the poverty rate for white children (7 percent) [47]. Median household income by race in Orleans Parish also reveals extreme racial disparities. In 2010, the median household income for whites was $57,031 well above the city average of $37,726. In contrast, the median household income level for African Americans was $30,167, well below the city average [48].

In short, modest gains in ethnoracial diversity have occurred in a context of intense and highly rigid patterns of racial residential segregation and high poverty rates. Despite the dramatic demographic reordering of neighborhoods since the disaster, the vestiges of segregation continue to permeate the socio-spatial makeup of the city and its neighborhoods. We can understand post-Katrina recovery and repopulation trends in the context of larger socio-spatial relations including enduring patterns of concentrated poverty and residential segregation. In the decades after 1970, the emergence of concentrated poverty and extreme segregation created a city with dramatic and troubling racial and class disparities [49,50,51,52]. While the entire city suffered from a low median household income and other social problems, the African American population suffered even more [53]. Moreover, the concentrated geographic pattern of poverty and residential segregation sharpened and intensified these problems among poor African Americans [54]. As a result, by the time Hurricane Katrina roared ashore, New Orleans had become a place sharply divided by race and class—a city where many poor black residents were geographically isolated from the rest of the population [55]. Because black neighborhoods experienced greater flooding and population loss after the deluge, these neighborhoods were more likely to diversify than areas that experienced less flooding and less population loss. Thus, post-Katrina repopulation rates reflect the legacies of past socio-spatial inequalities of race/ethnicity and class.

4.1. Resident Perceptions of Resilience: Insights from Ethnographic Field Observations and Interviews

Since the collapse of the Army Corps of Engineers levee system after Hurricane Katrina, the Lower Ninth Ward has attracted much attention from media and researchers interested in understanding resident interactions, reactions, and interpretations of the ongoing recovery and rebuilding process. Like other areas in the city, up to ten feet of floodwater engulfed the neighborhood—the worst degree of inundation in the city—and repopulation and physical rebuilding have been slow. By 2010, the neighborhood had only recovered approximately 20 percent of its pre-Katrina population (see Table 3). According to the 2010 census, only 2842 Lower Ninth Ward residents live in the neighborhood out of a pre-Katrina population of 14,008. During 2006 and 2007, many of the flooded homes in the Ninth Ward were bulldozed rather than repopulated by homeowners, a development that caused a 60 percent drop in the percentage of total housing units in the Lower Ninth Ward, from to 5601 housing units to 2039, a loss of 3562 units.

Through our interviews and ethnographic research, we have found that attachment to place is considered a strong indicator of neighborhood resilience but "attachment" has different meanings for residents. Several activities have been initiated by the Center for Sustainable Engagement and Development (CSED), local non-profit organizations, to inculcate a sense of place and foster an appreciation of the history and culture of the neighborhood. These activities include the "Food Action Plan" to promote urban farming and healthy food choices, the "Sustainability Series—Living with Water, Discussion and Site Tour", and the creation of Pam's Place, named in honor of CSED founder Pam Dashiell—a landscape depot, community garden and wetlands restoration exhibit. The aim of these and other community activities is to "restore the Lower Ninth Ward as a safe, environmentally just and economically vibrant community", according to one person we interviewed [56]. A strong sense of place is exemplified in many of the initiatives undertaken by CSED, such as supporting the construction of the Alfred Lawless School system that was flooded during the storm, the development of economic opportunities along the commercial corridors of the neighborhood, and assisting with efforts to establish a grocery store in the community.

For some residents, resilience is not a static descriptor of a neighborhood but is expressed in actions that assist with coastal restoration and wetlands regeneration. Resilience is also expressed, according to some residents, in efforts to eliminate blight, enhance walkability, and assist with the return of displaced residents. There is an important feedback loop working through these activities: place attachment generating activities foster a resilient neighborhood and a resilient neighborhood creates the conditions and context for engaged residents to generate strong attachments to community. There is an understanding among residents that place attachment is important for community resilience, and enhancing community resilience means strengthening people–place connections. Here bridging ties with extra-local organizations and federal and state government agencies (e.g., FEMA, Army Corps of Engineers, Louisiana Economic Development, *etc.*) interconnect with bonding ties among residents to shape and influence resident constructions of resilience. Bridging ties are not just networks for funneling resources but cross-scale interactions that facilitate information exchange and communication.

Table 3. Total population and ethnic/racial population for New Orleans and the Lower Ninth Ward, 2000–2010.

	2000	2010	Change 2000-2010
Orleans Parish:			
Total Population	484,676	343,829	−140,845 (−29%)
White	128,871 (27%)	104,770 (30%)	−24,101 (−37%)
African American	323,392 (67%)	204,866 (60%)	−118,526 (−19%)
Asian	11,007 (2.0%)	9,988 (3.0%)	−1,019 (−9.0%)
Hispanic	14,826	18,051	3,225 (21%)
Lower Ninth Ward:			
Total Population	14,008	2,842	−11,166 (−79.7%)
White	73 (0.5%)	50 (1.8%)	−23 (−31.5%)
African American	13,813 (98.6%)	2,713 (95.5%)	−11,100 (−80%)
Asian	5 (<0.1%)	2 (0.1%)	3 (−60%)
Hispanic	70 (0.5%)	42 (1.5%)	28 (−40%)

One meaning of resilience that radiates from the voices of residents is one in which current revitalization efforts are clearly and explicitly connected with collective memories and authentic accounts of the past. Resilience means the preservation of the neighborhood's history and knowledge, including knowledge of the surrounding wetland ecosystem. According to one resident we interviewed:

Actually, community restoration must be about residents becoming active stewards and caretakers of the environment. The culture of the wetlands, the waterways, and bayous must interact with the Lower Ninth Ward. If not, then culture will dissolve. An upscale development does mean community restoration or the preservation of culture. Cultures sit in neighborhoods and it is vital that we maintain that knowledge base—the history and culture of the community—in order to be resilient. And if history gets lost then we lose our community because there's nobody to say, that's not really what it was, no this is how it was [57]

The above quote suggests a meaning of resilience that is rooted in a conception of the Lower Ninth Ward as integrated with the nearby wetlands ecosystem. In particular, a source of hope and inspiration is the restoration of the Bayou Bienvenue, a 31,000 acre wetland ecosystem that has experience severe degradation from a combination of land and water management practices compounded by rising sea levels (see Figure 2). Understanding the trends of the ecosystem degradation is an important starting point for the ecosystem restoration and community resilience. The meaning is widely shared and is actively promoted by community organizations and non-profit organizations working in the neighborhood. According to the website of the CSED:

The health and well-being of the Lower Ninth Ward—its resilience as a community—is deeply intertwined with the Mississippi River on the south and the Bayou Bienvenue Wetland Triangle to the north. This new understanding continues to inform Coastal Sustainability and the critical role of wetlands throughout New Orleans, across Louisiana and along the Gulf Coast. [58]

In the quotes below, two residents reflect on the integration of the community with collective knowledge of the Bayou Bienvenue ecosystem that surrounds the neighborhood on the north, east, and south sides. For some residents, neighborhood identity is inseparable from the nearby marshes and bayous. Thus, resilience means not just repopulation and physical rebuilding but restoring the health and sustainability of the surrounding Bayou Bienvenue ecosystem. As one person told us,

People's attachment with the bayou makes the community less isolated than in the past. Since the storm, people are paying more attention to the bayou and wetlands and this has created an opportunity for the community to be more engaged. Also, when you look at the children and teenagers in the Lower Ninth Ward and also those who didn't grow up in that community but maybe came from other places, and can see how quickly the attachment to the natural environment becomes. People are now aware of how bayou can produce devastation. And this creates new opportunities for community preservation because people are more engaged with the bayou. We are not as isolated as we once were because people want to come to visit and learn from the Ninth Ward. [59]

Through various activities with residents and stakeholders, we are trying to make sure that the community and others continue to build a knowledge base about the area. While research and studies are important, there is still so much more to know about. I've seen in other parts of the country, how a body of water can make a community, or reinvent a community, whether it's a port city or a bay, and I think the bayou has that kind of possibility of what it can make to the lower nine and to New Orleans. [60]

Another interesting finding from our ethnographic field observations is the diversity of opinions and views of "sustainability" in discussions of resilience. In addition to typical neighborhood calls for repaired streets and code enforcement, Lower Ninth Ward residents now regularly lobby for sustainable development including structurally sound levees, bayou restoration, and environmentally sensitive construction. Yet tensions have emerged over whether "green" sustainability practices—e.g., radiant barrier installations, rooftop solar arrays, compact fluorescent light bulb replacements, and so on—are the best use of resources and time for creating and maintaining community resilience. Some residents have voiced concern that green sustainability practices, especially those espoused by extra-local non-profit organizations and foundations, are piecemeal and ad hoc. They "address individual homes but not the community as a whole", according to one resident [61]. Moreover, for some residents the emphasis on green sustainability diverts attention from more important social problems that plague the neighborhood such as intense poverty, chronic unemployment, poor schools, inadequate health care, and so on). As one resident told us,

It is great for the homes in the community to be green and conserve energy. But that is not enough. Restoration means restoring a sense of community not just the houses. This is an important issue that I think people do not understand. Green homes do not make a resilient community. [62]

Overall, Lower Ninth Ward residents have a common desire for resilient neighborhood institutions but they do not always agree on what resilience means. Like the scholarly literature that reveals mixed, diverse, and conflicting conceptions and indicators of resilience, residents have constructed a variety of meanings of resilience to understand the post-Katrina recovery process and to situate themselves as active agents in community rebuilding. In particular, residents routinely call for increased "sustainability" and "sustainable" engagement and development as a means to achieving great neighborhood recovery and resilience. At the same time, residents are mindful that resilience also means accepting and being aware of the continued vulnerability of the neighborhood and region to storms, floods, and other climate change induced disturbances. A resilient neighborhood, according to some neighbors we interviewed, is an "engaged" neighborhood that is active in taking steps to prepare for future hazards and disasters. Indeed, according to some residents, vulnerability awareness and disaster preparation are important mechanisms for fostering strong bonding links among people that reflects and reinforces a resilient community identity. These points suggest that for residents, resilience and vulnerability are not opposites or conflicting concepts but integral to everyday life and community organization.

5. CONCLUSIONS

Over the years, pundits and researchers have debated whether post-Katrina recovery and rebuilding efforts reflect a continuance of past socio-demographic disparities around the nexus of race and class or whether the region is experiencing a lessening of inequality and becoming more sustainable and resilient to disasters. A number of researchers and officials maintain that while the region has suffered greatly, public and private reinvestments have spurred a remarkable recovery. For Michael Hecht, president and CEO of Greater New Orleans Inc., New Orleans is a resilient city that "offers lessons for the rest of the United States. We are sitting in the middle of one of the greatest urban revivals of our lifetime. There is a real lesson in rebirth and recovery here for the entire nation. All you have to do is believe in miracles" [63]. Based on a comprehensive examination of the region in 2010 using 20 key indicators, the Brookings Institution observed that the New Orleans area "has become more 'resilient', with increased civic capacity and new systemic reforms, better positioning the metro area to adapt and transform its future" ([64], p. 2). Overall, according to the Brookings Institution, despite two post-Katrina blows—the Great Recession and the BP oil spill in the Gulf of Mexico—the "greater New Orleans is rebounding and, in some ways, doing so better than before" the Hurricane Katrina disaster ([64], p. 5).

Our analysis both challenges and supports these assertions. Pronouncements of resilience are are socially constructed and interconnect with questions of recovery indicators, pace, and scale. Moreover, our ethnographic field observations and interviews point to the multidimensionality of resilience in which resident perceptions show resilience to be heterogeneous, conflictual, and contested. The malleability and fluidity of resilience emanates from stakeholders' different and sometimes competing notions of their neighborhood's identity. These points dovetail with Amundsen's [65] recent call for researchers to consider the multidimensional, complex, and dynamic nature of resilience. Although some New Orleans residents may consider their neighborhoods to be resilient, the rate and magnitude of expected systemic global changes, especially climate change and increasing storm intensity and frequency, means that current resilience may be fleeting and ephemeral. Moreover, the current reality of rising sea levels and subsidence in southern Louisiana means that future resilience cannot be taken for granted and may in fact be an illusion.

Finally, our study suggests several directions for future research on the linkages among diversity, inequality, and post-disaster recovery. While scholars have pointed out that diversity is essential for absorbing disturbance and helping with regeneration and reorganization following disturbance, there is no guarantee that all types of diversity will result in positive outcomes. Moreover, Nelson *et al.* [32], Bell *et al.* [66], and Janssen [67] have suggested that diversity has trade-offs and can have costs, as well as benefits, in both ecological and social contexts. We have focused on one measure of diversity (ethnoracial diversity) and examined its interaction with inequality and damage to investigate post-disaster repopulation trends at the neighborhood scale. More research is needed to assess whether different types of diversity—e.g., institutions, socioeconomic status, land-uses, jobs, spatial distribution of housing, and so on—can produce different post-disaster outcomes and/or contribute to new risks and vulnerabilities. Moreover, we need more investigation on how the diversity of cross-scale interactions, networks, and flows of resources can shape and determine the pace and trajectory of post-disaster repopulation and recovery. It might be that some types of diversity may proliferate vulnerabilities to perturbation. The interactions among different types of diversity and different forms of inequality deserve careful attention.

ACKNOWLEDGEMENTS

This material is based upon work supported by the National Science Foundation under Award No. 0948993.

CONFLICTS OF INTEREST

The authors declare no conflict of interest.

REFERENCES AND NOTES

1. Naim Kapucu, Christopher V. Hawkins, and Fernando I. Rivera, eds. *Disaster Resiliency: Interdisciplinary Perspectives*. New York City: Routledge, 2012.

2. Mark Pelling. The Vulnerability of Cities: Natural Disasters and Social Resilience. London: Earthscan Publications, 2003.

3. Lance Gunderson. "Ecological and Human Community Resilience in Response to Natural Disasters." *Ecology and Society* 15, no. 2 art. 18. (2010). http://www.ecologyandsociety.org/vol15/iss2/art18/.

4. Gerard Hutter. "Organizing Social Resilience in the Context of Natural Hazards: A Research Note." *Natural Hazards* 67 (2011): 1–14.

5. Jianguo Wu, and Tong Wu. Ecological Resilience as a Foundation for Urban Design and Sustainability. In *Resilience in Ecology and Urban Design*. Edited by Steward Pickett, Mary Cadenasso, and Brian McGrath. Amsterdam: Springer, 2013, pp. 211–29.

6. Gernot Grabher. Adaptation at the Cost of Adaptability? In *Restructuring Networks in Post-Socialism. Legacies, Linkages and Localities*. Edited by Gernot Grabher, and David Stark. Oxford: Oxford University Press, 1997, pp. 107–34.

7. Gernot Grabher, and David Stark. "Organizing Diversity: Evolutionary Theory, Network Analysis and Post-Socialism." *Regional Studies* 31 (1997): 533–44.

8. Neil Adger. "Social and Ecological Resilience: Are They Related? " *Progress in Human Geography* 24 (2000): 347–64.

9. Neil Adger, Mick Kelly, and Alexandra Winkels. "Migration, Remittances, Livelihood Trajectories, and Social Resilience." *AMBIO* 31 (2002): 358–66.

10. Lawrence Vale, and Thomas J. Campanella, eds. *The Resilient City: How Modern Cities Recover from Disaster*. New York: Oxford University Press, 2005.

11. Carl Folke. "Resilience: The Emergence of a Perspective for Social-Ecological Systems Analyses." *Global Environmental Change* 16 (2006): 253–67.

12. Charles L. Redman. "Resilience Theory in Archaeology." *American Anthropologist* 107 (2005): 70–77.

13. Steward Pickett, Mary Cadenasso, and James Morgan Grove. "Resilient Cities: Meaning, Models, and Metaphor for Integrating the Ecological, Socio-economic, and Planning Realms." *Landscape and Urban Planning* 69 (2004): 369–84.

14. Ann S. Masten, and Jelena Obradovic. "Disaster Preparation and Recovery: Lessons from Research on Resilience in Human Development." *Ecology and Society* 13, no. 1 (2008): art. 9. http://www.ecologyandsociety.org/vol13/iss1/art9/.

15. Crawford S. Holling. "Resilience and Stability of Ecological Systems." *Annual Review of Ecology and Systematics* 4 (1973): 1–23.

16. Crawford S. Holling. The Resilience of Terrestrial Ecosystems: Local Surprise and Global Change. In *Sustainable Development of the Biosphere*. Edited by William Clark, and R. E. Munn. Cambridge: Cambridge University Press, 1986, pp. 292–317.

17. Crawford S. Holling. "Understanding the Complexity of Economic, Ecological, and Social Systems." *Ecosystems* 4 (2001): 390–405.

18. Lance Gunderson, and Crawford S. Holling, eds. *Panarchy: Understanding Transformations in Human and Natural Systems*. Washington: Island Press, 2002.

19. Millennium Ecosystem Assessment (MA). *Ecosystems and Human Well-Being: Current State and Trends*. Washington: Island Press, 2005.

20. Fridolin Simon Brand, and Kurt Jax. "Focusing the Meaning(s) of Resilience: Resilience as a Descriptive Concept and a Boundary Object." *Ecology and Society* 12, no. 1 (2007): art. 23. http://www.ecologyandsociety.org/vol12/iss1/art23/.

21. Thomas Dietz, Elinor Ostrom, and Patrick Stern. "The Struggle to Govern the Commons." *Science* 302 (2003): 1907–12.

22. Nancy Grimm, and Charles L. Redman. "Approaches to the Study of Urban Ecosystems: The Case of Central Arizona-Phoenix." *Urban Ecosystems* 7 (2004): 100–213.

23. Neil Adger, Terry P. Hughes, Carl Folke, Stephen R. Carpenter, and Johan Rockstrom. "Social-Ecological Resilience to Coastal Disasters." *Science* 309 (2005): 1036–39.

24. Fikret Berkes, Johan Colding, and Carl Folke, eds. Navigating Social-Ecological Systems: Management Practices and Social Mechanisms for Building Resilience. Cambridge: Cambridge University Press, 2003.

25. Rolf Pendalla, Kathryn A. Fosterb, and Margaret Cowella. "Resilience and Regions: Building Understanding of the Metaphor." *Cambridge Journal of Regions, Economy and Society* 3 (2010): 71–84.

26. Deborah Wallace, and Rodrick Wallace. "Urban Systems during Disasters: Factors for Resilience." *Ecology and Society* 13, no. 1 (2008): art. 18. http://www.ecologyandsociety.org/vol13/iss1/art18/.

27. Lenore Newman Ann Dale. "Network Structure, Diversity, and Proactive Resilience Building: A Response to Tompkins and Adger." *Ecology and Society* 10, no. 1 (2005): art. r2. http://www.ecologyandsociety.org/vol10/iss1/resp2/.

28. Kevin Fox Gotham, and Richard Campanella. "Coupled Vulnerability and Resilience: The Dynamics of Cross-scale Interactions in post-

Katrina New Orleans." *Ecology and Society* 16, no. 3 (2011): art. 12. http://dx.doi.org/10.5751/ES-04292-160312.

29. David R. Godschalk. "Urban Hazard Mitigation: Creating Resilient Cities." *Natural Hazards Review* 4 (2003): 136–43.

30. Bobbi Low, Elinor Ostrom, Carl Simon, and James Wilson. Redundancy and Diversity: Do They Influence Optimal Management? In *Navigating Social-Ecological Systems: Building Resilience for Complexity and Change*. Edited by Fikret Berkes, Johan Colding, and Carl Folke. Cambridge: Cambridge University Press, 2003, pp. 83–114.

31. U.S. Long Term Ecological Research Network (LTER). 2007, *The Decadal Plan for LTER: Integrative Science For Society And The Environment*. LTER Network Office Publication Series No. 24; Albuquerque, New Mexico, USA.

32. Margaret Nelson, Michelle Hegmon, Stephanie R. Kulow, Matthew A. Peeples, Keith W. Kintigh, and Ann P. Kinzig. "Resisting Diversity: A Long-Term Archaeological Study." *Ecology and Society* 16, no. 1 (2011): art. 25. http://www.ecologyandsociety.org/vol16/iss1/art25/.

33. Rodrick Wallace. "Social Disintegration and the Spread of AIDS: Thresholds for Propagation Along 'Sociogeographic' Networks." *Social Science and Medicine* 33 (1991): 1155–62.

34. Steve Carpenter, Brian Walker, Marty Anderies, and Nick Abel. "From Metaphor to Measurement: Resilience of What to What? " *Ecosystems* 4 (2001): 765–81.

35. Ash Amin. "Surviving the Turbulent Future." *Environment and Planning D: Society and Space* 31, no. 1 (2013): 140–56.

36. Gabriela Christmann, and Oliver Ibert. "Vulnerability and Resilience in a Socio-Spatial Perspective." *Raumforschung und Raumordnung* 70 (2012): 259–72.

37. Brookings Institution. *New Orleans after the Storm*. Washington: Brookings Institutioni, 2005.

38. Federal Emergency Management Agency (FEMA). "National Flood Insurance Program, 2012." Available online: http://www.fema .gov/national-flood-insurance-program (accessed on 10 September 2013).

39. Federal Emergency Management Agency (FEMA). *Substantial Damage Estimator (SDE): User Manual and Field Guidebook*; FEMA P-784 CD. Washington: Government Printing Office, 2013.

40. United States Department of Agriculture (USDA), Natural Resources Conservation Service. *Flood Damage Assessment Tools*; Washington: Government Printing Office, 2013.

41. Jack P. Gibbs, and William T. Martin. "Urbanization, Technology and the Division of Labor." *American Sociological Review* 27 (1962): 667–77.

42. John Iceland, and Erika Steinmetz. "The Effects of Using Census Block Groups Instead of Census Tracts When Examining Residential Housing Patterns." Unpublished manuscript. 2003. http://www.census.gov/hhes/www/housing/housing_patterns/pdf/unit_of_analysis.pdf.

43. Richard Campanella. "An Ethnic Geography of New Orleans." *Journal of American History* 94 (2007): 704–15.

44. Louise K. Comfort. "Cities at Risk: Hurricane Katrina and the Drowning of New Orleans." *Urban Affairs Review* 41 (2006): 501–16.

45. Richard Campanella. 'Two Centuries of Paradox': The Geography of New Orleans' African American Population, from Antebellum to Postdiluvian Times. Scheduled for publication in an as-yet untitled historical anthology on Hurricane Katrina; Edited by Randy Sparks. Baton Rouge: Louisiana State University Press, 2013, forthcoming.

46. John Logan, and Brian Stults. "The Persistence of Segregation in the Metropolis: New Findings from the 2010 Census." *Census Brief prepared for Project US 2010*, Available online: http://www.s4.brown.edu/us2010 (accessed on 19 November 2013).

47. Allison Plyer, and Elaine Ortiz. Poverty in Southeast Louisiana Post–Katrina. Greater New Orleans Community Data Center (GNOCDC). Available online: http://www.gnocdc.org/Poverty/index. html (accessed on November 2013).

48. Corporation for Enterprise Development (CFED) and Greater New Orleans Community Data Center. "Assets & Opportunity Profile: New Orleans." *CFED*, 14 August 2012. Available online: http://www.gnocdc.org/Assets&OpportunityProfile/index.html (accessed on 20 November 2012).

49. Arnold Hirsch. Simply a Matter of Black and White: The Transformation of Race and Politics in Twentieth Century New Orleans. In *Creole New Orleans: Race and Americanization*. Edited by Arnold R. Hirsch, and Joseph Logsdon. Baton Rouge: Louisiana State University Press, 1992.

50. Arnold Hirsch. New Orleans: Sunbelt in the Swamp. In *Sunbelt Cities: Politics and Growth since World War II*. Edited by Richard M. Bernard, and Bradley R. Rice. Austin: University of Texas PressNew Orleans, 1983.

51. Daphne Spain. "Race Relations and Residential Segregation in New Orleans: Two Centuries of Paradox." *The ANNALS of the American Academy of Political and Social Science* 441 (1979): 82–96.

52. Peter F. Burns, and Matthew O. Thomas. "A 'New' New Orleans? Understanding the Role of History and the State–local Relationship in the Recovery Process." *Journal of Urban Affairs* 30 (2008): 259–71.

53. Beverly Wright. Black New Orleans: The City that Care Forgot. In *In Search of the New South: The Black Urban Experience in the 1970s and 1980s*. Edited by Robert Bullard. Tuscaloosa: University of Alabama Press, 1989.

54. Christine Cook, and Mickey Lauria. "Urban Regeneration and Public Housing in New Orleans." *Urban Affairs Review* 30 (1995): 539–57.

55. Mickey Lauria, and Vern Baxter. "Residential Mortgage Foreclosure and Racial Transition in New Orleans." *Urban Affairs Review* 34 (1999): 757–86.

56. RRR. Resident of the Lower Ninth Ward. New Orleans, LA, USA. Interviewed by the authors, 15 June 2012. Tape recording.

57. BGG. Resident of the Lower Ninth Ward. New Orleans, LA, USA. Interviewed by the authors, 30 July 2012. Tape recording.

58. Center for Sustainable Engagement and Development. Natural Environment. Available online: http://www.sustainthenine.org/programs/natural-environment (accessed on 17 October 2013).

59. EGR. Resident of the Lower Ninth Ward. New Orleans, LA, USA. Interviewed by the authors, 1 October 2012. Tape recording.

60. GRK. Resident of the Lower Ninth Ward. New Orleans, LA, USA. Interviewed by the authors. 12 October 2012.

61. DBR. Resident of the Lower Ninth Ward. New Orleans, LA, USA. Interviewed by the authors. 2 September 2012.

62. KFG. Resident of the Lower Ninth Ward. New Orleans, LA, USA. Interviewed by the authors. 29 September 2012.

63. Ron Starner. "The New Orleans Miracle: Led by an Innovative Mayor, A Resilient City Offers Lessons for the Rest of the United States." *Site Selection: The Magazine of Corporate Real Estate*

Strategy and Area Economic Development, March 2011.http://www.siteselection.com/ssinsider/snapshot/new-orleans.cfm.

64. Amy Liu, and Allison Plyer. An Overview of Greater New Orleans: From Recovery to Transformation. In *The New Orleans Index at Five*. Washington: Brookings Institution and Greater New Orleans Community Data Center, 2010.

65. Helene Amundsen. "Illusions of Resilience? An Analysis of Community Responses to Change in Northern Norway." *Ecology and Society* 17, no. 4 (2012): art. 46. http://dx.doi.org/10.5751/ES-05142-170446.

66. Andrew Reid Bell, Nathan Lee Engle, and Maria Carmen Lemos. "How Does Diversity Matter? The Case of Brazilian River Basin Councils." *Ecology and Society* 16, no. 1 (2011): art. 42. http://www.ecologyandsociety.org/vol16/iss1/art42/.

67. Marco A. Janssen. "An Update on the Scholarly Networks on Resilience, Vulnerability, and Adaptation within the Human Dimensions of Global Environmental Change." *Ecology and Society* 12, no. 2 (2007): art. 9. http://www.ecologyandsociety.org/vol12/iss2/art9/.

5

WEB-BASED GIS FOR SPATIAL PATTERN DETECTION: APPLICATION TO MALARIA INCIDENCE IN VIETNAM

Thanh Quang Bui[1,*] and Hai Minh Pham[2]

VNU University of Science, 334 Nguyen Trai, Hanoi, Vietnam

ABSTRACT

Introduction

There is a great concern on how to build up an interoperable health information system of public health and health information technology within the development of public information and health surveillance programme. Technically, some major issues remain regarding to health data visualization, spatial processing of health data, health information dissemination, data sharing and the access of local communities to health information. In combination with GIS, we propose a technical framework for web-based health data visualization and spatial analysis.

Methods

Data was collected from open map-servers and geocoded by open data kit package and data geocoding tools. The Web-based system is designed based on Open-source frameworks and libraries. The system provides Web-based analyst tool for pattern detection through three spatial tests: Nearest neighbour, K function, and Spatial Autocorrelation.

Results

The result is a web-based GIS, through which end users can detect disease patterns via selecting area, spatial test parameters and contribute to managers and decision makers. The end users can be health practitioners, educators, local communities, health sector authorities and decision makers. This web-based system allows for the improvement of health related services to public sector users as well as citizens in a secure manner.

Conclusions

The combination of spatial statistics and web-based GIS can be a solution that helps empower health practitioners in direct and specific intersectional actions, thus provide for better analysis, control and decision-making.

Keywords: Web-based, Health pattern detection, Malaria

BACKGROUND

There is a great concern on how to build up an interoperable health information system of public health and health information technology within the development of public information and health surveillance programme. Technically, some major issues still remain in health sector that relate to health data visualization, spatial processing of health data, health information dissemination, data sharing and the access of local communities to health information (Boulos 2011; Highfield et al. 2011). Pfeiffer and Robinson (2008). Knowing of pattern of distribution enables health practitioners to understand underneath mechanism of diseases development over time and driving factors of those changes. Therefore, the identification and measurement of patterns is an important step in analysing geographic information (Graham et al. 2004; Elliott 2001; Pfeiffer and Robinson 2008).

In comprehensive analysis, a known pattern of disease distribution might reveal the processes underlying the spatial distributions of disease cases, through further analysing physical, environmental and social conditions of the study areas (Kienberger and Hagenlocher 2014; Odongo-Aginya et al. 2005). Spatial patterns can be classified as regular, random, or clustered. Meanwhile, analysis methods are grouped into 'Non-specific or clustering analysis' and 'specific or cluster detection analysis' (Elliott 2001; Wakefield and Kelsall 2000). McLafferty (2015); Pfeiffer and Robinson (2008) summarized spatial data analysis in epidemiology in their works. Gruebner et al. (2015) used spatial statistics methods to study geographic variability and to disclose patterns of psychological vulnerability and resilience factors in the aftermath of disasters. Goovaerts and Jacquez (2005) detected temporal changes of cancer rates using spatial statistics analysis. (Lin and Zhang 2004); Lin (2004) developed method for spatial clustering test of rare diseases.

Mostly, in epidemiological practices, health data is processed using spatial analysis methods (Auchincloss et al. 2012; Graham et al. 2004). In most cases, geo-referenced data is used, in combination with attribute data that describes the characteristics of disease locations. Based on that, visualization, exploration and modelling can be carried out to assist in decision making (Fig. 1). Despite acknowledging the importance of spatially explicit processes in determining disease risk, the use of spatial information beyond recording spatial location and mapping disease risk is rare (Jacquez 2000). However, the use of spatial information becomes popular with the integration of GIS and statistics package in processing health data (Delmelle et al. 2010) and in opening up data linked to health services (Shoultz 2015). In fact, technical aspect is not an obstacle for the widespread uses of GIS-based health applications, but non-technical aspects remain a major bottleneck (Carroll et al. 2014) that requires a comprehensive strategy for spatially enabled health services.

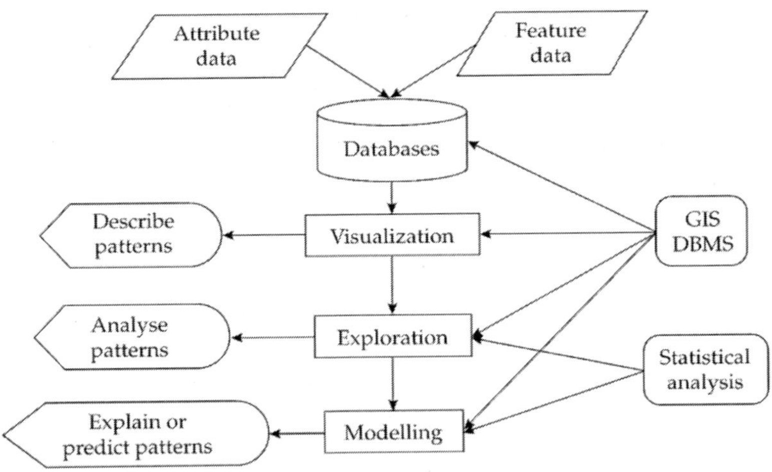

Fig. 1: Conceptual of epidemiology data analysis

Web-based information dissemination

The fast development of Internet and Web-based services helps health practitioners and communities to effectively and efficiently share health information and disseminate risks of diseases (Highfield et al. 2011), even though the services were limited in simple display and query tasks (Longley 2005). Recently, burgeoning efforts have been made to develop more active and dynamic systems and to make Web-based GIS more interactive for end users, such as statistical packages in health data processing and health information dissemination. Technically, geo-technology has enabled the health data sharing and analysis, but socially, it results in a major challenge in protecting patient geo-privacy while allowing analyses of spatial data (Hampton and Fitch 2010). The larger number of data sets become public, the more health information is exposed to communities. Recently, the opening up of health data leads to the development of filter techniques to mask out locations of patient's location information to preserve confidence. Although enormous efforts have been made, the instilment of sensitivity to geo-privacy and disclosure risk have been unsuccessful (Kounadi and Leitner 2014), particularly in the era of internet.

There is significant demand to exploit current geospatial information infrastructures in health applications (Granell et al. 2014). Researches have focused on the area of Web-based health information and more researchers (academic and non-academic) are using the web for data sharing, research and planning purposes (Lee 2010). Gao (2009) also emphasized the importance of comprehensive representation of heath information to users. Highfield et al. (2011) worked on interactive web-based mapping that provides health data easily accessible for a wide variety of research audiences. The authors urged that the system need to be as interactive as possible, allowing users to dynamically define their research parameters rather than being forced to accept pre-set queries. Boulos (2011) pointed out the application of open map services as Google, Openstreetmap to collect data from communities (crowd-sourcing) and discussed the benefits to get local communities involved in health surveillance. Delmelle and Zhu (2014) recently mentioned the state-of-the art web-based applications in epidemiology and produced an analytical module to map the outbreaks of dengue fever in Colombia, by using accelerated Kernel Density Estimation. Almost all mentioned researches assessed the available Web-based package for health information dissemination, but there are restrictions on spatial analytical capabilities (Delmelle and Zhu 2014) and lack of web-based pattern mapping tool, fostering multi-level participation to support identification of diseases clustering. Even though many web-based health applications dynamically generate view-only maps or interactive maps (Gao 2009) and allow users to study spatio-temporal distributions and evolution of patients (Hammond and Barzyk 2008). However, web-based health pattern detection tools were limited to some specialized researches, using specialized tools in particular a Web-based system (Mills 2008).

This paper focuses on creating web-based spatial processing tool that can be used to measure spatial patterns of the incidence over health data in Vietnam (in this case malaria samples were used). We develop three tools, namely nearest neighbour analysis, K-function, spatial autocorrelation to test the significances of clustering of point distribution. All points are treated in the distribution as if they are all the same they do not distinguish points by their attributes in the first two tools. The third uses aggregated data and takes into consideration both location and locational characteristics. These pattern detection tools can provide preliminary overview on how diseases clustering magnitude be found and will be useful for health practitioners.

METHODS

Malaria prevalence and pattern

The occurrence of malaria is closely related to environmental, climatic (Odongo-Aginya et al. 2005) and socio-economic conditions (Kienberger and Hagenlocher 2014). Malaria is still a major health problem in the world and Vietnam in particular. Though its prevalence has been reduced, there is great risk of diseases re-emergence to susceptible communities which live in the remote regions, especially those living in border areas between Vietnam and Laos and Cambodia (Hoang 2014) and in highland regions of Vietnam. Most of researches on malaria in Vietnam have focused on malaria epidemiology and achieved many results in prevention of malaria. But the disease is not adequately addressed due to lack of research on disease management, monitoring, detection and early treatment of malaria patients among local communities. A mechanism to share health information throughout communities is an important route to reduce national-wide public health issues (Hoang 2014). This research used malaria distribution as example data for spatial pattern detection tools.

Web-based spatial analysis tool

Spatial-analytical tools have increasingly been used with epidemiologic data to address the complex spatial nature of exposure-disease relationships. Management of the data is performed using geographic information systems (GIS) and database management systems, and is of relevance throughout the various phases of spatial data analysis. Many applications and GIS platforms demonstrate the application of spatial analysis in epidemiologic research, introducing analytic approaches which integrate GIS and standard statistical software tools to enhance epidemiologic assessments. They provide tools in visualization, exploration and modelling (Abdullahi et al. 2010; Dominkovics et al. 2011; Gao 2009). They cover approaches for identifying spatial clustering of disease, spatial interpolation and imputation methods, geographically weighted and land use regression, and basic spatial statistics useful for studying the spatial distribution of exposure and disease. Even though many Web-based health applications dynamically generate view-only maps or interactive maps, background processing and metadata are often missing (Gao et al. 2009).

In the literature review, several projects have focused on the use of representational state transfer (REST) architectural style and have proved it as an effective and efficient style in Web applications (Arakawa and Kido 2010; Dominkovics et al. 2011; Flemons and Guralnick 2007; Highfield et al. 2011; Kvilekval 2010). In general, clients interact directly with the exposed resources through a uniform interface, materialized basically through the combination of HTTP (hypertext transfer protocol), URI (uniform resource identifier) and standard formats such as HTML (hypertext markup language) and XML (extensible markup language).

System design

Web-based system provides fundamental tools for visualization, query and for spatial analysis in some cases. Users can interactively define study area and test whether observed distribution and known pattern are statistically different. It also provides an Information communication page. The communication component enables sharing of local/scientific knowledge and guidance of decision makers. The information system will be getting more and more enriched with analysed data by scientists, information verification and feedbacks from the users.

Figure 2 shows conceptual epidemiology data analysis capabilities in GIS, in which the authors propose an open-source framework to deal with pattern visualization and pattern analysis. The criteria for selection of systems embrace many aspects, such as Free and Open source software (FOSS), leverage open standards, hierarchical layering, spatial functionality, query layout, documentation and loading time. Among the two most popular open-source mapping servers UMN Mapserver and Geoserver, Geoserver was selected because of its extension that serves Web Processing Service. From the client side, there are many available frameworks that are sophisticated enough to compare with each other. This research employed GXP library, Openlayers, Extjs and GeoExt. It offers n-tier architecture consisting of: data layer, visualization and processing layer, and presentation layers.

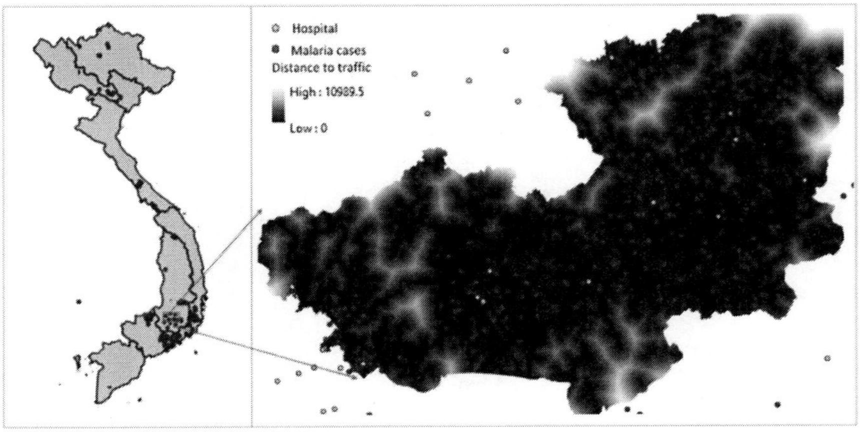

Fig. 2: Locations of reported malaria cases

CASE STUDY AND DATA COLLECTION

A case study was carried out in Lam Dong, Vietnam, a mountainous area in the central highland regions of Vietnam (Fig. 3). The study area is characterized by complex spatial variations in topography and climatic conditions. It is projected as one of the most vulnerable areas to malaria due to more complex topography, lower development index, difficult access to social services, and increasing pressure of population growth and its demand for production land. Furthermore, the combination of livelihoods of local communities (remote farming practices) and lack of mosquito control results in a significant increase in infected cases.

Data from the web For the case study, location of malaria cases were collected through internet-based open data, field data collection by open data kit (ODK) field survey and coordinate geocoding from statistical data. In spatial epidemiology, the most common form of data aggregation is sums of diseases cases in each area (Pfeiffer and Robinson 2008). For aggregated data, location is usually a generalisation of study area as centroids of administrative borders. The open malaria data from the last several years for all provinces of Vietnam was downloaded from (http://www.map.ox.ac.uk/ explorer/ http:// www.map.ox.ac.uk/explorer/). Data were filtered to remove unfit cases for the study.

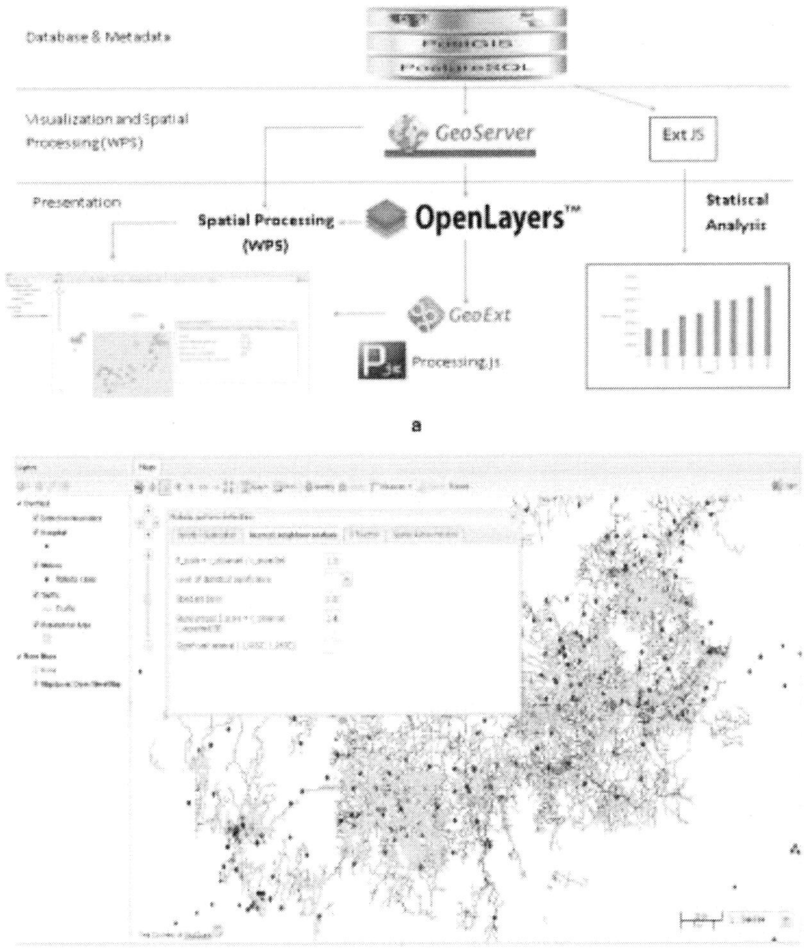

Fig. 3: Architectural framework **a** Architectural framework **b** WebGIS interface

Data geocoding Spatial location in maps can be defined by Geocoding tool that enriches a description of a location, most typically a postal address or place name, or matching column value of statistic table to attribute table of spatial data (shapefile). Geocoding provides position of points with a pair coordinate that facilitates spatial analysis in geographic information systems. This method is commonly used throughout health-related research, because health data often only contain low accurate position descriptions like district names and administrative units. (Dominkovics et al. 2011). Malaria cases in 2014 and first quarter of 2015 were collected mainly in the Lam Dong Department of Health and from the National Institute of Malariology Parasitology and Entomology (Fig. 3). Data collected should be taken into consideration with the extent of geographic areas in the study. Analysts often need to determine to what extent the areas surrounding the geographic objects of interest are to be included in their analysis.

Field data collection by ODK Additional malaria cases were added to the database using ODK. ODK is a free and open-source set of tools which help to provide mobile data collection solutions with GPS locations and images. ODK is being used to create decision support for technician to define diseases locations and for building multimedia-rich nature mapping tools. Using the ODK development interface, this research was able to develop a core template for the data collection survey (on mobile devices) that was then modified based on the specific data collection needs. Collected data was sent to Tomcat server, backed with PostgreSQL database. With updated points in the database, users can refresh browser for displaying map. Taking advantage of the integrated nature of the ODK Collect survey tool and ODK Aggregate server, surveys could quickly be modified, tested, and released for further testing or data collection.

RESULTS AND DISCUSSIONS

Density visualization

Based on test data, this tool performs a density measurement and visualization that provides a preliminary density visualization of diseases. Results depend on zoom level to show different magnitude of clustering. The processing service performs a heat map web processing service (WPS) service to compute a spatial density map to detect high concentration area of malaria cases. Malaria reported cases were stored in PostGIS and served by Geoserver. When the user makes a change to zoom level, request is sent back to server to recalculate the spatial map with changes in values according to the zoom level (Fig. 4).

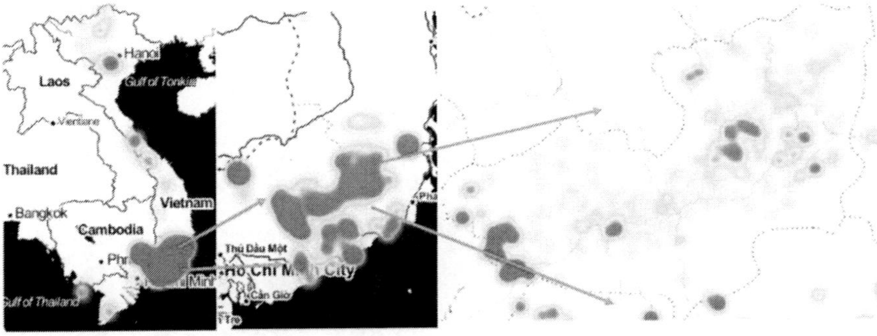

Fig. 4: Web-based density mapping

Nearest neighbour analysis

The test is based on comparing the observed average distances between nearest neighbouring points and those of a known pattern (Wong 2005) and is used to estimate the spatial proximity among events. User can draw a box around cluster of points to define boundary of study area, then select level of significance. The tool will calculate R scale, standard error, Z score through calling PostGIS function and compare it to significant interval to check whether the observed distribution is significant different from random pattern. The example showed that Z score is between (−1.96SE− >1.96SE) at 5 % significant level that means observed and random pattern are not statistically different (Fig. 5).

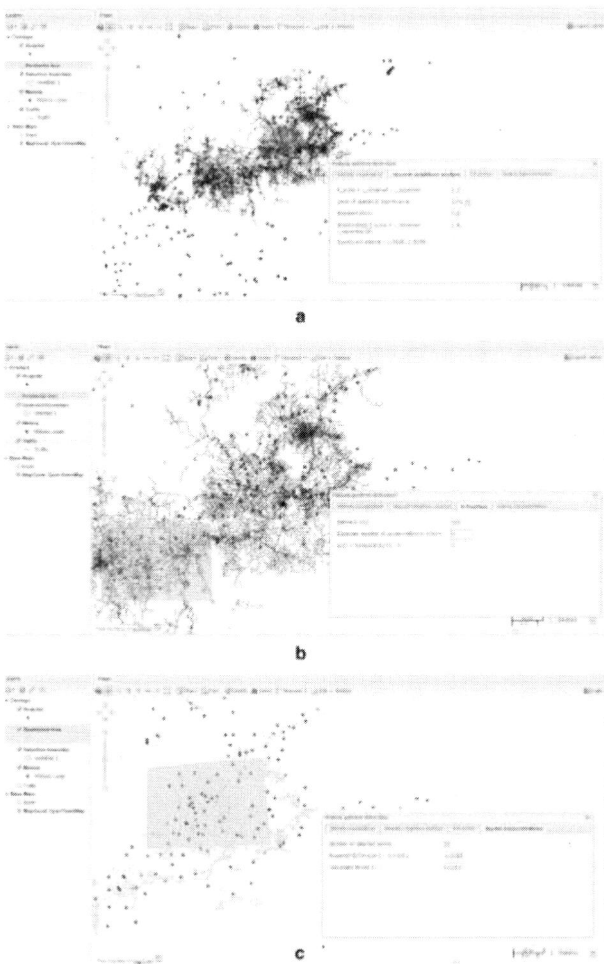

Fig. 5: Spatial tests **a** Nearest neighbour analysis **b** K-function **c** Spatial autocorrelation Moran's I

A function was built in PostGIS to calculate sum of nearest distances from selected points that are spatially within area of interest. The final sum of nearest distance is used for R test to check whether the observation is significantly different from random distribution.

- St_contain (geomA, geomB): This function selects all geomA (malaria points) that spatially are inside the bounding box of geomB (box that user draw).
- St_sum, St_Distance: The functions are used to calculate the shortest distances between malaria points and sum the nearest distances up.

With all measured parameters, this tool will calculate D statistic value and D_α for K–S test, by calling PostGIS function. The example showed that D statistics is greater than $D_\alpha = 5\ \%$ that means two distributions are significantly different in a statistical sense (Fig. 5a). As in the figure, we defined an area in Lam Dong province and the result was considered to be a non-random pattern at the $\alpha = 0.05$ level.

K-function

One of the limitations of the nearest neighbour distance method is that it uses only the nearest distance, as such it only considers only the shortest scales of variation. The K-function provides an estimate of spatial clustering over a wider range of scales. This higher-order analysis based on all the distances between events in the study area and assumes isotropy over the region. The tool starts with drawing a box as a bounding box for malaria point selection, and continues selecting a distance or spatial lag. For each point in the selected area, a buffer with radius equal to spatial lag is drawn and sum of numbers of points within each buffer is calculated.

We build a web-based tool where users can interactively set an area of interest and set value for initial distance (or spatial lag) h. We use functions in PostGIS to create buffer for each point and calculate the sum of numbers of point that are spatially within the buffers. Result are stored in an array for graph display and spatial homogenous test. The final sum of the nearest distance is used for R test to check whether the observation is significantly different from random distribution. $K(h)$ is then plotted against different values of h that describe paired comparison between observations and random patterns.

- St_buffer (geom, spatial lag): Thus function creates buffer for each point with radius defined by a spatial lag.
- St_contain (geomA, geomB): This function selects all geomA (malaria points) that spatially are inside the bounding box of geomB (box that user draw).

- St_sum, St_Distance: The functions are used to calculate the shortest distances between malaria points and sum the nearest distances up.

Spatial autocorrelation

Neither the nearest neighbour, nor the K-function take point's attributes into consideration. Different geographic locations rarely have identical characteristics. Because, this is aggregated data, in which one location might contain information on malaria incidences of an area (in this case, areas can be at commune or district level). The number of cases in an area can be used as weight of points in spatial analysis. Spatial autocorrelation is used in considering the effects of both distances between points and their attributes. With spatial autocorrelation, Moran's I index is used to measure the proximity of location and similarity of the locations. In this study we develop a PostGIS tool to measure Moran's I, in which data can be either continuous or count data.

Moran's I coefficient of autocorrelation quantifies the similarity of an outcome variable among points that are defined as spatially related. It measures the proximity of disease points and the similarity of the characteristics of those points (Wong 2005). The tool starts with drawing study area and then by selecting level of significance. It then calculates E(I), Z(I) value through spatially processing distances between points, through calling PostGIS function. This example showed that Z(I) lied between significant interval, so that the observed distribution is not statistically different from random pattern as described in Eq. (1).

$$Moran'sI = \frac{-\sum_{i=1}^{n}\sum_{j=1}^{n} w_{ij}(x_i-\overline{x})(x_j-\overline{x})}{s^2\sum_{i=1}^{n}\sum_{j=1}^{n} w_{ij}} \quad (1)$$

In which $s^2 = \frac{\sum_{i=1}^{n}(x_i-\overline{x})^2}{n}$, n = number of points.

- St_contain (geomA, geomB): This function selects all geomA (malaria points) that spatially are inside the bounding box of geomB (box that user draw).
- St_Distance: The functions are used to calculate the distances between malaria points and weight between point i and j is calculated as w_{ij} = 1/St_distance ($geom_i$, $geom_j$) w_{ii} is equal to 0.
- St_sum: Sum up attribute values of all points.

Expected E(I) is measured as E(I) = −1/(number of points −1). Cluster pattern, Random pattern and dispersed patterns are detected when calculated Moran's

I value is higher than E(I), equal to E(i) or small than E(I) respectively. In this example, with the calculated score was higher than expected E(I) of −0.185, the malaria points showed a slightly statistically significant departure from a random pattern. This was demonstrated by high score of 0.0213 [a bit higher than random E(I)].

The selection of proper geographic scale for clustering analysis affects the analysis results. Objects may be detected as clustered, dispersed or random depending on the actual scale (in this case zoom scale) we define before proceeding the analysis. In fact, data was geocoded using reported addresses. However, there were missing in documenting correct addresses, then they were aggregated into communes or even into districts. The effect is known as Modifiable aerial unit problem that can radically affect the analysis results.

Indeed, the examples describe three spatial detection tools that compare the observed distribution to random pattern of points, or more specifically pattern of diseases. These functionalities are powerful testing tool to convey message to and to support decision making. By zooming in and out, in combination with clustering tool, users define the boundary of study area and understand changes of analytical results that vary with zoom scale. Nearest neighbour can be used for the global test of clustering, it measures distances between points and compare these distances to known pattern. It has been extended (in comparison to Quadrat statistics) to accommodate second, third, and higher order neighbourhood definitions. The outcome results of this analysis vary depending on spatial scale, because some look clustered in small scale but seem dispersed at large scale. If boundary box size is too small, each polygon may contain a couple of points. On the other hand, if it is too large, each polygon contains many points. This measure does not take into account that different points may be different in how points are represented (due to their differences in characteristics).

The second analysis takes into consideration distances among points and results in plot, describing areas where patterns are clustered or dispersed. All spatial processing of three analyses are done in PostGIS and measured statistics are sent back to client. The third one measures the proximity of location in considering effects of locational characteristics. However, this test considers the population at risk is evenly distributed within the study area and correlation in all directions are to be the same (Moran 1950). This is a drawback of the Moran's I test. Through the Web based pattern detection tools, end users can interactively define (draw) area for measurement. The tools are three straightforward measurement of clustering and diversion that provide preliminary assessment to disease distribution. More spatial tests should be further incorporated.

USER EVALUATION

For test and evaluation, in order to improve the quality of the system, and to make the products more useful and usable, we have targeted group of epidemiologist and health practitioners in the National Institute of Malariology Parasitology and Entomology and the National Institute of Occupational and Environmental Health. The prototype was introduced and participants were guided to use the system to detect clustering in Lam dong province as demonstrated in the sections and feedbacks were reported. We collected 19 responses from the participants from both agencies. The questions focused on (1) whether the objectives of the application are relevant (2) user interface (3) are clustering methods useful (4) is the tool easy to convey a message to decision makers (5) what scale selection should be fit for clustering analysis. Values were given in range from 1 as worst to 10 as best. Overall assessment was shown in (Table 1).

Table 1: User evaluation

Content of survey / boxplot of feedbacks from 19 users (range from 1 as worst - 10 as best)	
• Objective of the application 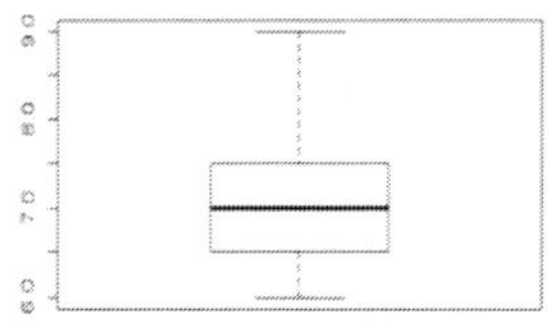	• User interface 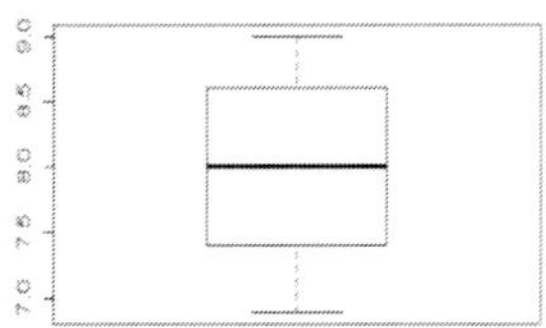
• Usefulness of clustering method 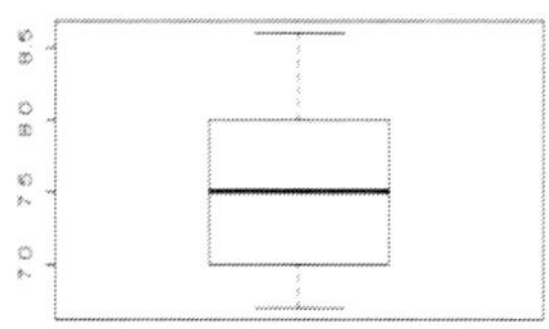	• Is tool easy to convey message to decision makers 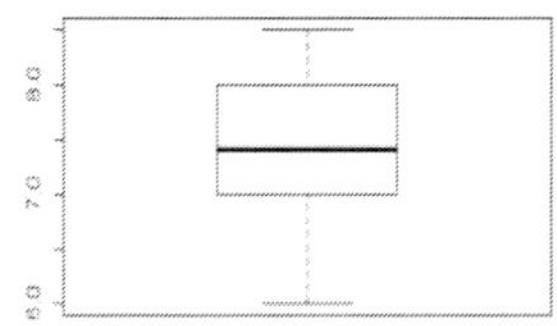

The results from questionnaire indicated the overall satisfaction of the surveyed group in the two organizations. System approach, user's interfaces, clustering usefulness and easiness to convey message, received upper moderate acceptance from participants, whereas the selection of scale for analysis was discussed among users. The survey revealed that an online clustering application had not been deployed in both agencies and there was relevant demand to have a system that support simple and fast analysis either on the field or indoor. On the other hand, the effects of scale on analysis are significant to conclude whether patterns are clustered/dispersed or randomly distributed. Based on clustering analysis, further studies on variety of driving forces of those pattern will be conducted. The users acknowledge the fundamentality of clustering in epidemiology and discussed on the need to have an interactive tool do the analysis. They commented on integrating administrative boundaries instead of interactively defining the boundary box. The survey results provided useful feedbacks from local administrations who are key people in decision making and local communities who are directly affected by disasters, and can be used for future research and improvement of the system.

CONCLUSIONS

In this paper, the WebGIS system was developed as a tool in promoting a user-friendly approach in detecting malaria pattern in Vietnam. Three pattern detection tools were developed as nearest neighbour, K-function, spatial autocorrelation to visually test health data distribution to known distribution or random distribution. Although, all points are treated in the distribution as if they are all the same they do not distinguish points by their attributes in the first two tools, the results might provide a visualized view on how health patterns are and where incidents are clustered. This system helps health professionals in analysis and decision-making tasks. The success of such pattern measurements mainly depends on the precise collection of malaria point samples and interactive identification of analysis boundary. This system is typically directed toward the widest possible end users (decision makers, managers, local communities) whose contributions may vary considerably. The combination of these benefits may result in direct, specific intersectional actions, and thus provide for better analysis, control and decision-making.

AUTHORS' CONTRIBUTIONS

All authors participated in the design, coding of the web-based system. Both authors read and approved the final manuscript.

COMPETING INTERESTS

The authors declare that they have no competing interests.

REFERENCES

1. Abdullahi FB, Lawal MM, Agushika JO (2010) Design and implementation of a Web-based GIS for public healthcare decision support system in Zaria metropolis. Int J Res Rev Appl Sci 4(4):435–439

2. Arakawa K, Kido N, Oshita K, Tomita M (2010) G-language genome analysis environment with REST and SOAP web service interfaces. Nucleic Acids Res 37:W700–W705

3. Auchincloss AH, Gebreab SY, Mair C, Roux AVD (2012) A review of spatial methods in epidemiology, 2000–2010. Annu Rev Public Health 33:107–122

4. Boulos K, Maged N, Resch B, Crowley DN, Breslin JG, Sohn G, Burtner R, Pike WA, Jezierski E, Chuang KYS (2011) Crowdsourcing, citizen sensing and sensor web technologies for public and environmental health surveillance and crisis management: trends, OGC standards and application examples. Int J Health Geogr 10(1):1–29

5. Carroll LN, Au AP, Detwiler LT, Fu TC, Painter IS, Abernethy NF (2014) Visualization and analytics tools for infectious disease epidemiology: a systematic review. J Biomed Inform 51:287–298

6. Delmelle E, Delmelle EC, Casas I, Barto T (2010) HELP: a GIS-based health exploratory analysis tool for practitioners. Appl Spat Anal Policy 4(2):113–137

7. Delmelle EM, Zhu H, Tang W, Casas I (2014) A web-based geospatial toolkit for the monitoring of dengue fever. Appl Geogr 52:144–152

8. Dominkovics P, Granell C, Pérez-Navarro A, Casals M, Orcau À, Caylà JA (2011) Development of spatial density maps based on geoprocessing web services: application to tuberculosis incidence in Barcelona, Spain. Int J Health Geogr 10(1):1–14

9. Elliott P, Wakefield J, Best N, Briggs D (eds) (2001) Spatial epidemiology: methods and applications. Oxford University Press, Oxford. doi:10.1093/acprof:oso/9780198515326.001.0001 **(Oxford Scholarship Online, 2009)**

10. Flemons P, Guralnick R, Krieger J, Ranipeta A, Neufeld D (2007) A web-based GIS tool for exploring the world's biodiversity: the global biodiversity information facility mapping and analysis portal application (GBIFMAPA). Ecol Inform 2:49–69

11. Gao S, Mioc D, Yi X, Anton F, Oldfield E, Coleman DJ (2009) Towards web-based representation and processing of health information. Int J Health Geogr 8(1):1–14

12. Goovaerts P, Jacquez GM (2005) Detection of temporal changes in the spatial distribution of cancer rates using local Moran's I and geostatistically simulated spatial neutral models. J Geogr Syst 7(1):137–159

13. Graham AJ, Atkinson PM, Danson FM (2004) Spatial analysis for epidemiology. Acta Trop 91(3):219–225

14. Granell C, Fernández ÓB, Díaz L (2014) Geospatial information infrastructures to address spatial needs in health: collaboration, challenges and opportunities. Future Gener Comput Syst 31:213–222

15. Gruebner O, Lowe SR, Sampson L, Galea S (2015) The geography of post-disaster mental health: spatial patterning of psychological vulnerability and resilience factors in New York City after Hurricane Sandy. Int J Health Geogr 14:1–6

16. Hammond D, Barzyk TCK, Zartarian V, Schultz B (2008) Application of GIS mapping tools to prioritize community air pollution issues. Epidemiology 19(6):S173

17. Hampton KH, Fitch MK, Allshouse William B, Doherty Irene A, Gesink Dionne C, Leone Peter A, Serre Marc L, Miller William C (2010) Mapping health data: improved privacy protection with donut method geomasking. Am J Epidemiol 172(9):1062–1069

18. Highfield L, Arthasarnprasit J, Ottenweller CA, Dasprez A (2011) Interactive web-based mapping: bridging technology and data for health. Int J Health Geogr 10:69

19. Hoang H (2014) Malaria situation and evaluation of malaria control interventions in several border communes of Huong Hoa district, Quang Tri province PhD in Community Medicine. Hue College of Medicine and Pharmacy, Hue

20. Jacquez GM (2000) Spatial analysis in epidemiology: nascent science or a failure of GIS. J Geogr Syst 2:91–97

21. Kienberger S, Hagenlocher M (2014) Spatial-explicit modeling of social vulnerability to malaria in East Africa. Int J Health Geogrs 13:29

22. Kounadi O, Leitner M (2014) Why does geoprivacy matter? The scientific publication of confidential data presented on maps. J Empir Res Hum Res Ethics 9(4):34–45

23. Kvilekval KFD, Obara B, Singh A, Manjunath BS (2010) Bisque: a platform for bioimage analysis and management. Bioinformatics 26(4):544–552

24. Lee BK (2010) Epidemiologic research and web 2.0—the user-driven web. Epidemiology 21(6):760–763

25. Lin G (2004) Comparing spatial clustering tests based on rare to common spatial events. Comput Environ Urban Syst 28(6):691–699

26. Lin G, Zhang T (2004) A method for testing low-value spatial clustering for rare diseases. Acta Trop 91(3):279–289

27. Longley PA, Goodchild MF, Maguire DJ, Rhind DW (2005) Geographical information systems and science, 2nd edn. Wiley, New York, pp 291–292

28. McLafferty S (2015) Disease cluster detection methods: recent developments and public health implications. Ann GIS 21(2):127–133

29. Mills JW, Curtis A (2008) Geospatial approaches for disease risk communication in marginalized communities. Prog Comm Health Partnersh Res Educ Action 2(1):61–72

30. Moran PAP (1950) Notes on continuous stochastic phenomena. Biometrika 37:17–23

31. Odongo-Aginya E, Ssegwanyi G, Kategere P, Vuzi PC (2005) Relationship between malaria infection intensity and rainfall pattern in Entebbe peninsula, Uganda. Afr Health Sci 5(3):238–245

32. Pfeiffer DU, Robinson TPSM, Stevens KB, Rogers DJ, Clements AC (2008) Spatial analysis in epidemiology. Oxford University Press, Oxford

33. Shoultz J (2015) How to build and ArcGIS open data site for health and human services, 2015 health and human services GIS conference. ESRI, Atlanta, Georgia (2015)

34. Wakefield JC, Kelsall JE, Morris SE (2000) Clustering, cluster detection and spatial variation in risk. In: Elliott P, Wakefield JC, Best NG, Briggs DJ (eds) Spatial epidemiology—methods and applications. Oxford University Press, Oxford, pp 128–152

35. Wong DWS, Lee J (2005) Statistical analysis of geographic information with ArcView GIS and ArcGIS. Wiley, New York

36. http://www.map.ox.ac.uk/explorer/ Providing open access data online to advance malaria research and control. Accessed 19 Nov 2015

6

EVALUATION OF SPATIAL ANALYSIS APPLICATION FOR URBAN EMERGENCY MANAGEMENT

Rifaat Abdalla

Department of Hydrographic Surveying, Faculty of Maritime Studies, King, Abdulaziz University, P. O. Box 80401, Jeddah 21589, Kingdom of Saudi Arabia

ABSTRACT

Background

This paper provides multidisciplinary scope to the utilization of geospatial data frameworks for urban disaster management with accentuation on particular events. The emergency management events presented in this review are universally known and represent high risk for different parts of the world.

Results

The discussion starts with addressing the application issues related to how spatial analysis can be used intending to disaster management operations by characterizing its ease of use and impediments in managing the inquiries of vulnerability and hazard assessment. It also highlights best practices for the approaches to integrating spatial data for hazard mapping and risk perception.

Conclusions

The goal of this study is to give conceptual coverage to appropriate solutions for emergency preparedness and response, using spatial analysis and GIS. The paper emphasized that among different issues that may confront the use of spatial analysis, is the accuracy of data and time of processing, in addition to collective coordination of stakeholders working in the field. The findings of this research conclude that a challenge to possible risk reduction is furnishing disaster managers with access to information and methodologies that may help them in analyzing, evaluating and mapping hazard models.

Keywords: GIS, Emergency management, Urban centers, Spatial analysis, Environmental modeling

BACKGROUND

GIS applications in Disaster Management are progressively turning into a necessary component of disaster and emergency management activities in many parts of the world. The time considerations are extremely critical in emergency management operations. Emergency Managers are required to take significant decisions, promptly to provide fast response to extreme situations. The spatial dimension of geospatial data makes it exceptionally critical for decision-makers in the different phases of emergency management operations. It is important for policy makers to have the right information at the ideal time exhibited off base models to permit them to react, arrange or moderate catastrophes. The temporal nature of disasters does not allow emergency managers to gather the critical data, in a timely, in many situations. As such, more often, pre-arranged disaster management scenarios are utilized Becerra-Fernandez et al. (2008). GIS Technology is capable of filling up the gap of perception and investigation of simulating emergency scenarios showing various situations and their temporal attributes. This permits disaster managers to have access to sufficient data stored in spatial databases and exhibited in a PC created maps or intuitive models Miura et al. (2007). GIS can be exceptionally useful to make well-thought counter disaster response patterns, which can address the overall population. It is a helpful tool in disaster management planning, tabletop activities, and a fundamental element of Emergency Operations Centers (EOC) (ESRI 1999). GIS gives a component to perception and demonstrating of primary data different levels of details and for various regions after a disaster strikes (ESRI 1999). This provides a user-driven approach, which envelops the phases of disaster management, to bolster the procedure of improved primary leadership and builds the level of inclusion of every group of workforce related exercises and systematic methods (Smirnov et al. 2006).

Discussion about disaster management is tending to the issues of preparedness. This is a crucial part of disaster and emergency management and can assume an indispensable part if contingency activities, which gets to be vital. The convenience of GIS as a decision support system is in helping disaster managers and emergency first responders to falls in the following:

1. Risk and Threads Assessment.
2. What-if scenario modeling.
3. Maintaining situational awareness.
4. Allocation of Resources and documenting disruptions.
5. Alerting and notification of communities.
6. Minimizing vital service disruptions during the response stage.

Saadatseresht et al. (2009) have presented the factors above are especially of significance from spatial analysis point of view In an emergency management situation. It showed that spatial analysis can be performed for emergency management evacuation operations, in order to relocated population at risk for a safer location, this is usually a complicated process, dense population. Anjum et al. (2011) indicated that it is important to use utilize the state-of-the-art of spatial analysis tools for emergency planning operations, related to evacuation of masses, during extreme events. A major challenge for using spatial analysis as a part of search and rescue plans is in deciding the evacuation process to relocate the evacuees to a safer place. This indicates that supporting the choice of where and from which street every evacuee ought to go is an essential factor in the best utilization of spatial analysis capabilities (Cova and Church 1997). Several factors are involved in determining the efficiency of utilizing the process of spatial analysis for disaster management. To adequately accomplish the point of spatial investigation in crisis administration operations, a few goals are brought into thought and fulfilled at the same time through this paper. These objectives are (a) how a decision-maker can utilize the effectiveness of spatial analysis for prioritizing important decisions, during an emergency. (b) What are essential capacities that spatial analysis can help with amid disaster management cycle in the ten chosen disaster themes focused in this paper? (c) How decision-makers could better actualize spatial analysis process as a significant aspect of their everyday operations.

Successful disaster management calls for including multi-modal decision-making competencies; that includes aggregates at all levels of relief and response, notwithstanding total relief endeavors that address the origin of vulnerability. Morrow (1999) the vulnerability of group is correctly credited to the socioeconomic variables that influence the group, i.e., the directly affected, whether expanded or diminished as a consequence of the socio-economic well-being of a community, as it identifies with their everyday practice (ESRI 1999). Emergency Management planners, policy makers, risk analysts and first responders usually attempt to characterize and find high-risk factors utilizing Community Vulnerability Maps, consolidating this information into GIS frameworks, and for this, spatial analysis is essential (Kumar 2013).

EMERGENCY MANAGEMENT OPERATIONS

Comprehensive Emergency Management (CEM) is a concept that ensures the effectiveness of all aspects of emergency management by anticipating, minimizing the risks introduced by various emergencies, by preparing for emergency situations, and by helping in recovering from an emergencies. This approach is systematically addressed by Gordon (2002), where he provided a framework for comprehensive approach for dealing with risks.

According to Bullock et al. (2006), the impact of disaster can impact every community, state or authorities within the proximity of the event. As such, there is no person immune to the impact of disasters. According to Mileti (1999) many disaster losses are to some degree predictable, which make them manageable to a certain extent. Effective processes in disaster management help to reduce devastation and high costs at both the local and regional levels. A comprehensive emergency management system is composed of the interaction of policies and procedures, as well as the institutional and financial mechanisms, to constitute community-based approach to disaster risk management (Carter 1991).

URBAN EMERGENCY MANAGEMENT

This part covers a critical review on the use of spatial analysis in some urban emergency management situations. It provides an inside and out scope of the work cited in this regard to in such manner that it highlights the process of enhanced decision-making process. The extent of the scope will concentrate on the most important progressions in the utilization of spatial analysis methods for emergency management in urban situations.

Spatial analysis applications in natural hazards

Earthquakes and humanitarian coordination

The literature on GIS and humanitarian coordination has started by first looking at the different approaches in which GIS can be utilized for effective coordination. Regardless of the way that GIS has been predominately seen, within the disaster management community, as a cartographic tool, an approach to managing initial analysis and visualization, or an electronic navigational system, this does not attractively depict the best way of GIS utilization in humanitarian assistance (Currion 2006). There are numerous potential utilizations of GIS for humanitarian aid. For instance, the usage of enhancement, which is the use of cutting edge GIS calculations to take care of an outline issue, can be utilized to discover reasonable areas for clearing. For example, a support investigation, for analysis of spatial relationships using GIS as a tool, can be employed to gauge vulnerability to various hazards based on proximity.

In 2005, a Complex Humanitarian Emergencies Study by Verjee (2005) drew from contextual analyses and examples in innovative progression to format the potential GIS applications for humanitarian emergencies, which were:

1. Mapping and Cartography (Land use Mapping, Infrastructure Mapping, Demographic Mapping, Logistics, and Sustainability).
2. Outreach, Media and Communications (Public Access to Information, Reporting, Program Assessment, News Coverage).
3. Modeling and Simulation for Disaster Scenarios (Practice, drills, and exercises, Data information flow, planning for contingencies).
4. Environmental Management and Planning (Planning, Yield Cultivation, resources assessment).
5. Risk and Hazard Management (seismic analysis, site selection and planning, and water level estimation and mitigation).
6. Vulnerability Analysis and Assessment (Early Warning frameworks for the dry season, desertification and starvation, Epidemics modeling and Tsunami Planning).
7. Risk Reduction ('problem areas' distinguishing proof and relief programming).
8. Response Policies and Organizational Management (administration, planning, and training).

Table 1 is demonstrating the capabilities of GIS in this situation. In spite of the fact that there are various applications of spatial analysis as a GIS technique, they all share an ultimate target, which is to which is to exploit the situational the situational awareness to all areas taking an interest so fundamental concerns can be perceived and after that together achieved.

Table 1: Applications of spatial analysis in humanitarian coordination

GIS application	Description of application	Tools	Application examples
Queries and measurements			
Queries	Does not change the database or produce new data but reveals information	Scatter plots, residuals, and structured/standard query language	To determine water sources near potential IDP settlements
Measurements	Numerical interrogations of GIS, which make an analysis of spatial data	Makeup distance, area, length, shape, aspect of spatial data	To estimate the total area needed to set up IDP settlements To determine the length between food distribution points
Transformations			
Transformations	Creates new data from existing data	Buffer analysis	Used to comply with Sphere humanitarian standards. For example, can be used to ensure that toilets are

GIS application	Description of application	Tools	Application examples
			within 50 meters of shelters
Spatial interpolation	A transformation analysis method used for intelligently guessing the value of the discreet object	Thiessen polygons, inverse-distance weighting, and Kriging	To predict infection rates in IDP settlements To estimate flow rates of well to large aquifer
Optimization			
Point optimization	Determines optimal locations amongst nodes of a network	Network analysis	To select location for critical infrastructure such as food storage areas
Route optimization	Determines optimal routes amongst nodes of a network	Network analysis	To access accessibility of road networks to potential IDP settlements
Path optimization	Solves are routing problems by minimizing friction value and does not require a network	Network analyst	To determine paths for utility infrastructure for IDP settlements
Analysis			

GIS application	Description of application	Tools	Application examples
Geostatistical analysis	Used to measure geographic distributions, identify pattern and clusters, and analyze geographic relationships	Geostatistical analysis	To determine historical rainfall averages over a particular area, which could be applied to potential IDP locations
Centroid analysis	Used to identify trends over various phenomena in a set period	Spatial analyst	To determine trends in infectious diseases and household income levels within IDP settlements
Pattern analysis	Used to identify distribution points. Can determine whether the point is random, clustered or dispersed	Spatial analyst	To identify links between water wells and infectious diseases
Relationship analysis	Used to determine the relationship between various geographic phenomena	ArcGIS 3D analyst, ArcGIS spatial analyst and ArcGIS network analyst	To assist in increasing the efficiency of daily operation of IDP settlements
Geovisualization	Simple three dimensions (3D) and high-	Google earth, ESRI's arcexplorer, and microsoft virtual	Allows a layperson to understand the

GIS application	Description of application	Tools	Application examples
	resolution mapping are used globally to access a variety of information	earth	unfolding of humanitarian emergency

A late analysis by Eveleigh et al. (2007) and Al-Ahmadi et al. (2014) has utilized spatial analysis for earthquake disaster studies. The adopted approach recognizes that within the scope of humanitarian assistance "GIS innovation is battling with how to address complex issues that require the modeling of rapidly changing dynamic phenomena, feature, behavior, data and. They concluded that there is a high potential for GIS-based assessment models to give the leap forward expected to address the random way of humanitarian emergencies.

Bally et al. (2005) presented the use of remote sensing for Humanitarian Aid, showing that the utilization of remote detecting and GIS permitted 200,000 IDPS to be migrated to longer-term settlements that had a renewable water source and with improvement potential in regards to sanitation, farming, and even hydropower. Another powerful GIS application used to support humanitarian emergencies was The Global Connection Project, which included Carnegie Mellon University, NASA, Google and National Geographic, contributing to the relief planning for October 8, 2005, South Asian earthquake and tsunami. In this project, GIS was utilized to gain and convey high-resolution imagery from Digital Globe's Quickbird.

Wild fire

ESRI (1999) has shown an approach to depicting a rapidly spreading fire event precisely; spatial analysis can be utilized to recognize high-risk fire zones and set up buffer zones for evacuation. Notwithstanding the determination of high-risk regions, spatial analysis can be combined with statistical analysis as a verification method for the specifying areas of final damage assessment, in addition to deciding to provide visual models for highly impacted areas, according to Goodchild (2006). Lentile et al. (2006) gave direction by distinguishing potential layers that can be utilized for urban fire identification. The initial step was to employ scope and longitude directions to plot the different flames (based upon a decision of lightning or human-ignited fire) during a particular period. Fire information may seem, by all accounts, to be situated inside waterways. However, this is mainly a reason for adjusting buffer zones to give some slack to such errors. The process of connecting attributes information and present four analytical techniques for simulation and visualization out of control fire. In spite of the fact that their emphasis particularly on human-brought ablaze catastrophes, in proposing the four prescribed alternatives for finish urban fire examination:

a. the territory influenced

b. temporal expand

c. spatial extend

d. probability

The urban fire hazard is hard to avert. Notwithstanding, through the recognizable specification of the high-risk zones, the recurrence of flame can be minimized. Jaiswal et al. (2002) have demonstrated that GIS when joined with satellite imagery, can be useful in identifying high-hazard regions within given vicinity and restrict the fire spread and thus minimize the impact. Jaiswal et al. (2002) have also examined the utilization of ArcGIS for this idea, declaring that the mix of topographic foundation data and remote sensing for vegetation mapping can make a precise estimation of high-risk fire territories utilized for moderation and reaction purposes. In Jaiswal et al. (2002) different layers of vegetation, slope, proximity to settlements, and distance from roads were made to provide an indication about high-risk fire regions. After this data was plotted, buffer zones of 1000, 2000, 3000, and 4000 m surrounding the high-risk zones were plotted to extend the distinctive levels of danger. Although they have investigated a particular instance of India, the concept of using GIS spatial analysis consolidated with satellite imagery for distinguishing areas prone to high-risk of fire hazard has demonstrated the adequacy of GIS as a tool for urban disaster management. If GIS can be utilized to model and simulate high-risk fire zones with buffers, which gives benchmark understanding that GIS could likewise be used to show damage assessment models using different software and different data layers, regardless of the geographic location.

Pradhan et al. (2007) utilized GIS examination to decide fire susceptibility, using a "vector spatial database" with GIS and consolidated with topographic information, fuel information, base overview focuses, and maps. This took into account figuring variables, which were then changed over to a raster grid, recognizing 112 cells inside the fire events. A frequency-based proportion approach was used to characterize the "connections between hotspot areas and the components in the study area". The challenges, notwithstanding, were in processing "a significant amount of data". The conclusion is drawn from Pradhan et al. (2007) on the utilization of such projections for fire risk mapping and mitigation was quite compelling. In foreseeing fire susceptibility when utilizing frequency analysis, the prescribed results were recommended to be used with alert, according to Pradhan et al. (2007). It was suggested that the analysis approaches their examination is used fundamentally amid fire event, which proposes mapping fire-influenced zones instead of driving toward the relief bit of fire disaster management process.

Floods

Correia et al. (1998) demonstrated that GIS had been seen as a successful tool to organize and visualize data from different sources on far-reaching floodplain administration. As a part of this overall approach to manage floodplain management, it is crucial to have the ability to predict the aftereffects of different situations as to flooded regions and related regions at risk. Morrow (1999) discussed the hydrologic and water controlled zones accept a crucial part, and there is much to get in uniting these exhibiting capacities in a GIS system. The perspective of the using Intergraph GIS with IDRISI GIS provided an effective way in dealing with flood emergencies in both 2D and 3D. Using multidimensional modeling usually extended the flexibility of using GIS as an instrument for flood modeling. Gogoaşe Nistoran et al. (2016) have shown the effectiveness of spatial analysis using GIS for modeling flood inundation as a result of dam-break.

The role of GIS in Flood Disaster Management was analyzed by Cova (1999), through the perspective of Comprehensive Emergency Management (CEM) and its four phases: mitigation, preparedness, response and recovery. In the wake of a disaster, GIS is getting the chance to be vital in supporting damage assessment, evaluation, and cost estimation for development. In the aftermath of a catastrophe, GIS is a valuable tool in supporting cost evaluation and rebuilding. Abbas et al. (2009) proposed a GIS-based contemplate regarding the change of surge showing and representation for Allahabad Sadar Sub-District (India). This joins the framework, the methodology/approach that planned to research the degree for spatial analysis application for a rapid response. The flood affected zones have been recognized, and their positions are checked, where the GIS handiness has been manhandled to get the spatial information for the fruitful calamity organization for surge affected reaches. The adopted approach has helped in recognizing issues that may upgrade the present practices of emergency management organizations. The approach gives a suitable and quick fundamental authority instrument for snappy response to emergencies if used appropriately, which along these lines would help in minimizing loss of life and property. Al-Sabhan et al. (2003) proposed a GIS-construct study, in light of the change of flood levels and representation. This consolidates the arrangement, the investigated the present status of progressing hydrological models used for flood modeling and risk mitigation. It indicated how electronic systems could overcome a bit of the obstruction of existing structures. While hydrological GIS-based models are open, they are ineffectually suited to the consistent application and are frequently not primarily consolidated with spatial datasets.

Buchele et al. (2006) and Chen et al. (2009) discussed a modern approach for integrated flood risk assessment. In light of the setting of a more relative examination of different flood risk assessment models, for mapping, in the midst of extreme situations. The ampleness of synchronous and in-house proprietary methods using was analyzed by Chen and Zhan (2008). The study used an operator-based technique to model movement streams at the level of individual vehicles and examines the total practices of modeling and visualization of moving objects, during an emergency. De Silva (2000) presented a model Spatial Decision Support System (SDSS) which was normal for credibility making blueprints for emergency mapping, where response operations using spatial information dealing with and representation points of confinement in a GIS. It interfaces together with the geospatial part of the spatial analysis section is given by the GIS. The SDSS, so that gives a detailed spatial information of flood zone extension and involved layers.

Moreri et al. (2008) proposed an approach to manage making an internet-based Geographic Information Systems (WebGIS) application, which would reinforce people living in flood zones, which may at one point be unprotected in light of their closeness to the stream and the adequacy of the flooding. Zerger and Wealands (2004) showed that spatially quick hydrodynamic flood models could expect an essential part in average danger peril reducing. A key element of these models that make them suitable for risk exhibiting is the capacity to give time-blueprint immersion data about the onset, length, and embarking to an emergency situation. Such data can be the start for region utilize orchestrating, for mapping, for clearing directing, and for finding sensible crisis organization to give a couple of representations hazard responses. To address these confinements, a structure has been made that interfaces, with emergency response team with a GIS-based decision support system.

Dust storms

Dust Storms are otherwise called Sand Storms; it represents one of the common hazards with a broad range of environmental impacts. During an event of a stand storm, it affects human health in various ways. Sandstorms are a critical reason for car crashes and cause air transportation delays. Goudie (2008a, b) discussed the products during the process of stand storm eruption. It presents fine particles, salts and chemicals (counting herbicides) into the environment, with a suite of health effects, including respiratory complaints as well as different serious illnesses. Dust storms can transport allergens including microscopic organisms and growths, in this manner affect human health. Spatial Analysis can be exceptionally successful in displaying and representation the degree and the effect of sandstorms. Specifically, we can utilize GIS to give the accompanying capacities in managing dust storms disaster management.

The recent developments in global warming and climate change have prompted increased activity of sand storms in various parts of the word. Numerous researchers including Goudie (2008a, b; Xu et al. 2006) have dealt with the examination of sandstorms events and its impact on the land surface, utilizing GIS and Remote Sensing. Goudie (2009), discussed the first methodology relies upon the investigation of weather station information and representation of the spread of particulate matter in particular space in association with Dry Mid Temperature and Sub-Dry Temperature, particularly in the desert or semi-desert or zones. Measurable investigates exhibit that the event of sand–dust storms relate to a high degree of wind speed, which thus is firmly identified with land surface components; then again, a significant relationship between rain event and other atmospheric elements, for example, precipitation and temperature were not watched. This is notwithstanding the part of vegetation cover, which has been unequivocally connected to dust storms.

Health hazards

According to Cioccio and Michael (2007) Emergency management of health impacts, specifically focus on the vulnerable population; and access to medical services; GIS technology is capable in extreme heat attacks, by providing the degree and application for spatially analyzing the distribution of services and its relation to the population at risk. Despite all that, the literature that covers the use of GIS for health impact is somewhat limited. Many requests for the use of GIS in health focuses on the methodology, and the practical applications to the domains of vulnerable population, health care facilities distribution, and emergency shelters distribution. These three themes could be linked to the census and traffic information to provide more detailed spatial models, when dealing with this hazard.

Sharma et al. (2008) pointed out that one of the key utilizations of GIS in pandemic modeling and simulation is to encourage access to health services by inhabitants who live in and around the security area of a mass gathering or a social event. This will be achieved by outlining an application GIS to help health authorities in the planning and implementation of emergency medical response, with an emphasis on improving support of vulnerable population, including:

a. Ensuring continuous routine for health services amid times of restricted access to a security area;

b. Ensuring evacuation procedures for medical emergencies that are non-event related;

c. Providing timely evacuation and health care in the event of mass causality incident.

This can be accomplished by outlining a mapping tool to locate vulnerable community members inside the affected zone, if there should arise an occurrence of a pending natural or technological disaster, for example, a heat wave or power outage. Becerra-Fernandez et al. (2008), explained that GIS could be used to for specifying access and evacuation routes, for approaching or in progress emergency or disaster management events. Goals may incorporate shelters, schools or other predefined destinations outside of the security zone Chandana et al. (2007). The key support of GIS in a pandemic episode can be through the utilization of GIS intending to general public health issues, particularly, to characterize its uses and restrictions in managing the inquiries of describing the vulnerable population. GIS supports advanced intervention operations, for example, Roland Daley et al. (2015) have highlighted some of these issues as following:

1. Choosing sites for community flu centers and vaccination stations.
2. Monitoring and assessing effect of vaccination centers and stations.
3. Canceling public events, and gatherings.
4. Closing schools, meetings and gatherings.
5. Restricting utilization of public transportation frameworks.
6. Identifying potential groups quarantine and isolation facilities.
7. Enforcing people to follow group or individual isolates.

Spatial analysis applications in technological hazards

Infrastructure disruption and malfunction

Cova and Church (1997) and Cimellaro (2016) discussed an approach for purposely recognizing neighborhoods that may go up against transportation challenges during an emergency evacuation. A description of this nature offers an interesting approach to manage assessing group of defenselessness in regions subject to advanced dynamic risks of uncertain spatial impact (e.g., hazardous spills on roadways). A heuristic estimation is delineated which can be usable for conveying useful, the excellent answer for this model in a GIS setting, as it was associated with a study region.

Camps (1993) presented a new computerized risk management framework for use by less experienced risk management personnel who to reduce the likelihood and seriousness of accidents. The framework, which was developed, is suitable for use in oil, gas, or chemical processing sites. It joins scientific models and calculation tools for accident simulation and building a database that incorporates accidents scenarios and response plans. It can likewise be utilized as a part of an emergency situation to decide favored approaches to find external assistance.

Spatial analysis applications in manmade hazards

Mass gathering and civil unrest

Numerous sorts of mass gathering and the concentration of population change participating in such events may vary, depends on the nature of the event, its location and the time and season of the event. For instance, civil demonstrations, outdoor rock concerts, and a football match are typical examples where there is clear variation in the density of population attending these events. According to McDonald (2008), these occasions regularly, don't draw in one sort of participants. Therefore, risks might be connected with weather related sickness, harmful impacts of medications, or injury because of members attempting to draw near to the stage. Bradler et al. (2008), concluded that political events, for example, political parties conventions might have several risks, associated with. This incorporate trauma or toxic impacts of depression related to a political protest or terrorism-related incidents. Becerra-Fernandez et al. (2008) have indicated that GIS spatial analysis is valid in this applications, as it provides:

1. Specifying the dissemination of individuals around the event proximity.
2. Analyzing the scope and approach for mapping evacuation if there should be an occurrence of an emergency.
3. Determining the positions and movement of law enforcement in the field.
4. Analyzing the pattern of development of masses.
5. Supporting effective decision-making on evacuation and response to an emergency situation.

Terrorism

Kwan and Lee (2005) have shown that the terrorist attacks on the World Trade Center (WTC) in New York City and the Pentagon on September 11, 2001, has not quite recently impacted multi-level structures in an urban center. They have also influenced by their surroundings at the street level in ways that reduced the time limits for the speed of emergency response. The capacity of using progressing 3D GIS for the headway and execution of GIS-based intelligent emergency response systems. The fact was at urging a quick emergency response to terrorist attacks on multi-level structures (e.g., multi-story office structures). A system design and a framework data show that facilitates the ground transportation capabilities with the inside courses inside multi-level structures into a protected 3D GIS was examined. Issues of using adaptable representation stages were also discussed especially the prerequisite for the remote and versatile response plan. Critical decision support functionalities were moreover considered with particular reference to the utilization of framework based most restricted way computations. A test use of expected 3D structure data shows a GIS database for a nearby study area was demonstrated by Kwan and Lee (2005). The study indicates that reaction delay inside multi-level structures can be any longer than deferrals caused on the ground transportation framework, have the potential for impressively decreasing these postponements.

Johnson (2003) demonstrated that in times of crisis, the disaster managers have the necessary commitment in regards to quickly and adequately managing any situation that may happen. An adjusted GIS application was delivered engaging a brief based examination of a catastrophe occasion facilitated with the centralization of masses distinguished correctly to the room level. The GIS Emergency Management System (GEMS) application is an astute structure to be utilized as a part of the Emergency Operation Center (EOC) to help the heading of the response. On the off chance that a calamity needs to happen, the intervention and recovery attempts could be at initially focused on the most fundamental areas of the greatest convergence of people.

CHALLENGES AND TRENDS

Goodchild (2006) indicated that various events, including the Indian Ocean Tsunami of 2005, the Hurricanes of the 2005 season, and the 7/7 and 9/11 terrorist's attacks, have made each one of us seriously aware of the shortcoming of the modern society. Knowing the chronicled record of the events and where such situations have happened, notwithstanding the geographic limits of their effects are apparently essential, primarily when consolidated with data on human population and distribution, along with other spatially circulated wonders that might apply to reaction and recuperation (Abdalla et al. 2014). Regardless, GIS and spatial technologies that gather, analyze, and take into account visualization of such data, using advanced geomatics technologies in the form of GIS, remote sensing, GPS, and Photogrammetry. These technologies are unmistakably indispensable in all parts of the disaster management cycle, from protection, response, and recovery through acknowledgment to the reply and conceivable recovery. GIS give the preface to evaluating and mapping hazards, from evacuation planning to delivery to shelters, to routes planning to rehabilitation and restoration. Abdalla (2015) describes it as it likewise allows for choosing ranges where human population is well on the way to have been influenced by a disaster, and for allocating assets amid the recuperation procedure, among various other irreplaceable and crucial assignments that the GIS brings.

One of the distinct challenges in the utilization of GIS for urban disaster management is the location dependency of the event, or what is known as the geographic interdependence of the event. The proximity of the event can prompt many complexities in deciding the location, space and temporal parameters of the event. Various episodes in similar areas can bring about falling or raising impacts among the diversely affected entities.

As to the utility of Spatial Analysis in coordination approaches, the status of the spatial analysis confront noteworthy insufficiencies that can be condensed by the absence of some standard methods in a few districts and additionally the unstructured conventions that are being used when managing complex situations of disasters in various parts of the world. Some portion of this can be additionally ascribed to the absence of preparing in utilizing GIS frameworks, which can at times cause specialists on call for act wastefully when managing calamities. Spatial Analysis can bolster viable preparing for disaster management professionals to manage complex circumstances. More critically, the advocated requirement for geospatial information in a few parts of the world requires compelling access to worldwide SDI utilizing advanced access protocols and to strengthen the advancement of techniques and strategies for effective decision-making. This can be ascribed to the absence of interoperability in data exchange and processes standardization.

The decision-making process confronts a few difficulties on the lack of common operating picture for efficient policy implementation process. Decision-makers require to access spatial models actually in the form of real-time data feeds from planning teams, from field observers and remotely gathered information. This supports the viable access to operations areas monitoring devices, and as well it enhances the output of emergency management decision making.

SUMMARY

The spatial analysis gives powerful means in managing risk and in dealing with hazard mapping and assessment. It is likewise compelling in providing visual models that assist decision-makers with utilizing these advances adequately. However, spatial analysis can't be satisfied without getting the exact information, defining proficient strategies and securing viable execution of HR required in Emergency Management.

The extent of this exploration is challenging, with an endeavor to cover a few risks that may cause disasters in urban regions. It is essential to deal adequately with emergency management in any of its four stages, i.e., preparedness, mitigation, response, and recovery. The discussion of the detailed application of spatial analysis overviewing the state-of-the-art application the technology has raised a few issues identified with the development of the utilization of spatial advances in disaster management, specific questions are:

1. The co-locality of an impact as a result of a series of events may require more progressed spatial analysis answers for giving details about the extent of damage, the cost of harm, the distribution of vulnerable population, the indicators of vulnerability and the mean for a response.
2. Issues with data and information frameworks of interoperability are essential in giving an available prearrangement through spatial analysis.
3. Health related emergencies are more complex to analyze spatially because of issues identified with private access to patient's data, also the difficulty of covering various scales of events in the limited temporal timeframe.
4. Although the present GIS systems contribute to advanced spatial analysis capabilities; yet new methodologies for investigation, representation, and integration are required to offer additional means for support to urban disaster and emergency management community.

ABBREVIATIONS

GIS: geospatial information systems

GEMS: GIS emergency management system

EOC: emergency operations centers

ESRI: Environmental Systems Research Institute

SDSS: spatial decision support system

SDI: spatial data infrastructure

GPS: global positioning system

3D: three dimensional

HR: human resources

ACKNOWLEDGEMENTS

This project was funded by the Deanship of Scientific Research (DSR) at King Abdulaziz University, Jeddah, under Grant No. G-290/980/1434. The author, therefore, acknowledge with thanks DSR for technical and financial support.

COMPETING INTERESTS

The author declare that he has no competing interests.

AUTHORS' INFORMATION

Dr. Rifaat Abdalla is Associate Professor of Geospatial Information Science with King Abdulaziz University, Jeddah. During the past ten years, he served as Defence Scientist with Defence Research and Development Canada; the Federal Government of Canada research arm in Defense and Security Research. Also, he was Adjunct Professor with the graduate program in Disaster and Emergency Management, York University, Toronto; he also served as a contract faculty with the Geomatics Program, Ryerson University, Toronto. Dr. Abdalla received his Ph.D. from York University, in 2006. Dr. Abdalla has published extensively in the field of Web-based GIS for Disaster Management and has received many recognitions and awards, including the 2007 ESRI Best Scientific Paper, presented by the American Society for Photogrammetry and Remote Sensing (ASPRS) for papers published in Photogrammetric Engineering and Remote Sensing Journal and the Best Paper Award of 2009 Conference on Geo-Web Services, held in Cancun, Mexico.

AVAILABILITY OF DATA AND MATERIALS

The material used in this is paper is in form of published articles and they are all listed in the reference section.

FUNDING

This research was part of a project funded by King Abdulaziz University, Deanship of Scientific Research under Grant Number G-290/980/1434.

REFERENCES

1. Abbas S et al (2009) GIS-based disaster management: a case study for Allahabad Sadar subdistrict (India). Int J Manag Environ Qual 20(1):33–51
2. Abdalla R (2015) Strategic framework for advancing the utility of GeoICT in emergency preparedness. In: Proceedings of the 2nd international conference on information and communication technologies for disaster management (ICT-DM). IEEE, Rene
3. Abdalla R, Elawad Y et al (2014) A GIS-supported fuzzy-set approach for flood risk assessment. Can Water Resour J Rev Can Ressour Hydr 39(1):3–14
4. Al-Ahmadi K, Abdullah A-A, Linda S (2014) A spatial statistical analysis of the occurrence of earthquakes along the Red Sea floor spreading: clusters of seismicity. Arab J Geosci 7(7):2893–2904
5. Al-Sabhan W, Mulligan M et al (2003) A real-time hydrological model for flood prediction using GIS and the WWW. Comput Environ Urban Syst 27:9–32
6. Anjum M, Rana M et al (2011) GIS-based emergency management scenario for urban petroleum storage tanks. In: Proceedings of the 2011 international workshop on multi-platform/multi-sensor remote sensing and mapping
7. Bally P et al (2005) Remote sensing and humanitarian aid: a life-saving combination. Bull Eur Space Agency (ESA) 122:37–41
8. Becerra-Fernandez I, Prietula M et al (2008) Design and development of a virtual emergency operations center for disaster management research, training, and discovery. In: Proceedings of the 41st annual Hawaii international conference on system sciences
9. Bradler D, Kangasharju J et al (2008) Systematic first response use case evaluation. In: Second international conference on pervasive computing technologies for healthcare, 2008. PervasiveHealth 2008

10. Buchele B et al (2006) Flood-risk mapping: contributions towards an enhanced assessment of extreme events and associated risks. Nat Hazards Earth Syst Sci 6:485–503

11. Bullock JA, Haddow GD, Coppola DP (2006) Introduction to homeland security. Butterworth-Heinemann, Waltham

12. Camps R (1993) Controlling the danger zone. New Sci 138(4):16–17

13. Carter WN (1991) Disaster management: a disaster manager's handbook. Asian Development Bank, Manila

14. Chandana S, Leung H et al (2007) Disaster management model based on modified fuzzy cognitive maps. In: IEEE international conference on systems, man and cybernetics, 2007. ISIC

15. Chen X, Zhan FB (2008) Agent-based modelling and simulation of urban evacuation: elative effectiveness of simultaneous and staged evacuation strategies. J Oper Res Soc 59:25–33

16. Chen J, Hill AA et al (2009) A GIS-based model for urban flood inundation. J Hydrol 373:184–192

17. Cimellaro GP (2016) A comprehensive methodology for the evaluation of infrastructure interdependencies. Urban resilience for emergency response and recovery: fundamental concepts and applications. Springer, Cham, pp 139–223

18. Cioccio L, Michael EJ (2007) Hazard or disaster: tourism management for the inevitable in Northeast Victoria. Tour Manag 28(1):1–11

19. Correia FN et al (1998) Coupling GIS with hydrologic and hydraulic flood modelling. Water Resour Manag 12:229–249

20. Cova TJ (1999) GIS in emergency management. In: Longley MFGPA, Maguire DJ, Rhind DW (eds) Geographical information systems: principles, techniques, applications, and management. Wiley, New York, pp 845–858

21. Cova TJ, Church RL (1997) Modelling community evacuation vulnerability using GIS. Int J Geogr Inform Sci 11(8):163–784

22. Currion P (2006) Emergency capacity building project information technology and requirements. Assessment Report Global. 2006, Emergency Capacity Building Project. I. W. Group. London, p 34

23. de Silva FN (2000) Integrating simulation modelling and GIS spatial support systems for evacuation planning. J Oper Res Soc 51(4):423–430

24. ESRI (1999) GIS for emergency management. ESRI White Paper

25. Eveleigh TJ, Mazzuchi TA et al (2007) Spatially-aware systems engineering design modeling applied to natural hazard vulnerability assessment. Syst Eng 10(3):16–24

26. Gogoaşe Nistoran DE, Gheorghe Popovici DA et al (2016) GIS for dam-break flooding. In: Boştenaru Dan M, Crăciun C (eds) Study area: Bicaz-Izvorul Muntelui (Romania). Space and time visualisation. Springer, Cham, pp 253–280

27. Goodchild MF (2006) GIS and disasters: planning for catastrophe. Comput Environ Urban Syst 30(2006):227–229

28. Gordon JA (2002) Comprehensive emergency management for local governments: demystifying emergency planning. Rothstein Publishing, Brookfield, Connecticut

29. Goudie AS (2008a) Dust storms: recent developments. J Environ Manag 90:89–94

30. Goudie AS (2008b) The history and nature of wind erosion in deserts. Annu Rev Earth Planet Sci 36:97–119

31. Goudie AS (2009) Global deserts and their geomorphological diversity. In: Parsons AJ, Abrahams AD (eds) Geomorphology of desert environments, 2nd edn. Springer, New York, pp 9–20

32. Jaiswal RK, Mukherjee S et al (2002) Forest fire risk zone mapping from satellite imagery and GIS. Int J Appl Earth Obs Geoinform 4(1):1–10

33. Johnson K (2003) GIS emergency management for the University of Redlands. In: ESRI international user conference

34. Kumar A (2013) Natural hazards of the Arabian Peninsula: their causes and possible remediation. In: Earth system processes and disaster management. Society of Earth Scientists Series, pp 155–180

35. Kwan M, Lee J (2005) Emergency response after 9/11: the potential of real-time 3D GIS for quick emergency response in micro-spatial environments. Comput Environ Urban Syst 29(2):93–113

36. Lentile LB, Smith FW, Shepperd WD (2006) Influence of topography and forest structure on patterns of mixed severity fire in ponderosa pine forests of the South Dakota Black Hills, USA. Int J Wildland Fire 15(4):557–566

37. McDonald R (2008) New considerations for security compliance, reliability and business continuity. In: Rural electric power conference, 2008 IEEE

38. Mileti D (1999) Disasters by design: a reassessment of natural hazards in the United States. Joseph Henry Press, Washington, DC

39. Miura H, Yamazaki F et al (2007) Identification of damaged areas due to the 2006 Central Java, Indonesia earthquake using satellite optical images. In: Urban remote sensing joint event, 2007

40. Morrow B (1999) Identifying and mapping community vulnerability. Disasters 23(1):1–18

41. Moreri KK et al (2008) Web based geographic information systems for a flood emergency evacuation. In: 3rd International ISCRAM China workshop, Harbin

42. Pradhan B, Suliman MD et al (2007) Forest fire susceptibility and risk mapping using remote sensing and geographical information systems (GIS). Disaster Prev Manag 16(3):344–356

43. Roland Daley FF, Orly J, Abdalla R (2015) GIS for pandemic zoning: application of Brampton, Ontario, Canada. Comput Sci Inform Technol 3(5):149–158

44. Saadatseresht M, Mansourian A et al (2009) Evacuation planning using multiobjective evolutionary optimization approach. Eur J Oper Res 198(2009):305–314

45. Sharma RK, Kumar BS et al (2008) SAR for disaster management. IEEE Aerosp Electron Syst Mag 23(6):4–9

46. Smirnov A, Pashkin M et al (2006) Context-based disaster management support. In: IEEE workshop on distributed intelligent systems: collective intelligence and its applications, 2006. DIS 2006

47. Verjee F (2005) The application of geomatics in complex humanitarian emergencies. J Humanit Assist

48. Xu X, Levy JK et al (2006) An investigation of sand–dust storm events and land surface characteristics in China using NOAA NDVI data. Glob Planet Change 52:182–196

49. Zerger A, Wealands S (2004) Beyond modelling: linking models with GIS for flood risk management. Nat Hazards 33:191–208

7

MONITORING AND MODELING OF URBAN SPRAWL THROUGH REMOTE SENSING AND GIS IN KUALA LUMPUR, MALAYSIA

Mukesh Singh Boori[1,2,3], Maik Netzband[2], Komal Choudhary[1] and Vít Voženílek[1]

[1]Palacky University Olomouc, 17. listopadu 50, 771 46 Olomouc, Czech Republic

[2]Geographical Institute, Ruhr University Bochum, Bochum, Germany

[3]Samara State Aerospace University, Samara, Russia.

ABSTRACT

Introduction

Kuala Lumpur is currently experiencing a large-scale urban transformation. This study explores a multi-buffer ring method to investigate the spatial and temporal dynamics of urban growth in the Kuala Lumpur metropolitan area by using Geographical Information Systems (GIS) and remote sensing imageries from1989 to 2014.

Methods

The first Landsat images from 1989, 2001, and 2014 were classified with maximum likelihood classification to produce land cover maps and identify four types of land cover: urban/built-up, agriculture, forest, and water. Then, the area coverage for all the land use types at different points in time were measured and combined with distance from the city center. Later on, urbanization densities from the city center to the outside for every 1-km distance from 1 to 50 km are calculated.

Results

City density on the distance of 5, 8, 15, 20, and 30 km was very high in the years 1989 and 2014. Temporally, urban expansion shows fast and slow growth stages with high-speed growth shifting to the east side of the city.

Conclusions

In general, the population, traffic conditions, industrialization, and policy are the major factors thatinfluenced the urban expansion

Keywords: Urbanization, Landsat, Change detection, Remote Sensing, GIS

INTRODUCTION

Kuala Lumpur is one of the most important historical, cultural, industrial, and commercial cities in Malaysia as well as one of the most important cities in Asia. It has a booming economy with dynamic growth. During the last three decades, growth in population, industry, and economic activities has been strongly concentrated in the Kuala Lumpur metropolitan area. Formed in the 1850s, the Federal Territory of Kuala Lumpur is the national capital of Malaysia. The city has an estimated population of 1.627 million (2010 census). Malaysia's rich and diverse urban history developed primarily from the second half of the twentieth century when it shifted from a largely rural to a largely urban population. Following independence in 1957 and the formation of Malaysia in 1963, Kuala Lumpur became the administrative center and took center stage in national development. From the mid-1980s, Kuala Lumpur underwent policy shifts, notably privatization and economic liberalization (Malaysia 1991). Major development and supporting policy shifts were undertaken, and within two decades, there were major transformations and dramatic landscape changes in Kuala Lumpur (Saboori et al. 2012).

In the mid-1980s, there was a gradual but obvious change in the nature of the Malaysian economy, as was examined earlier. In 1987, for instance, the agricultural sector share in GDP was exceeded by the share of the manufacturing sector. In 1990, the agricultural share was 19.4 % compared with 26.6 % for the manufacturing sector (Saboori et al. 2012). Thus, there has been a change in the structure of Malaysia from an agricultural base to a manufacturing base since 1987. This, indeed, marks a milestone in the country's transition towards an industrializing economy (Malaysia 1991). It has often been argued that the process of urbanization and urban growth in the third world has taken place with the expansion of the industrial sector. That is true to a large extent in that, even without a substantial industrial base prior to 1957, Malaysia had experienced a large increase in its urban population. However, with the growth of manufacturing since 1957, the impact on the process of urbanization and urban growth has been most tangible as the employment opportunities generated were located in the urban areas (Salari et al. 2013; Boori and Vozenilek 2014). These issues make the necessity to the study of urbanization in Kuala Lumpur.

The Kuala Lumpur Structure Plan 2020 indicates that the population base is set to increase from 1.4 million to 2.2 million over the next 20 years, and that optimizing limited land resources will be a priority. The new Kuala Lumpur 2020 Strategy Plan was undertaken in the early 2000s to revise the 1984 Kuala Lumpur Structure Plan and to cope with the unprecedented growth and changes in the urban landscape. It is made compulsory to study Kuala Lumpur urban growth. Urbanization has been traditionally understood using the frameworks of demography, urban planning, urban infrastructure, transportation, affordable housing, and urban development (Yin et al. 2011; Boori et al 2014a). Urban growth is characterized by significant gains in urban/built-up areas at the expense of green or open spaces. In this study, the urban/built-up land cover class is defined as the areas dominated by buildings, asphalt, and concrete, including residential, commercial, industrial, and transportation space (Yin et al. 2011; Boori and Vozenilek 2014; Salari et al. 2013). Other types of urban land use, such as golf courses and urban green parks are not included in the urban/built-up class. Urbanization studies are useful to planners who seek to avoid the irreversible and cumulative effects of urban growth and to optimize the allocation of urban services (Dewan and Yamaguchi 2009a; Barnsley and Barr 1996; Boori et al. 2014b, Thinh et al. 2013). Furthermore, such information is indispensable for the assessment and evaluation of sustainable urban and environmental planning strategies (Wu and Hao 2012; Alphan 2003; Jensen and Im 2007). Land use planning is an integral part of the viability and liveability of a metropolitan area because it affects decisions about where and how we live. Over the last few decades, government agencies and local councils in different countries have been continually developing and adopting land use planning strategies to contain and manage urban growth (Yin et al. 2011). For example, "Smart Growth" in the USA and the "compact city" and "intensification" in Europe were devised and adopted to regulate and contain urban sprawl. There are few other studies such as Yagoub (2004) who focused his attention on Al-Ain in the United Arab Emirates. In another study, Al-Awadhi (2007) used available maps, aerial photographs, and satellite images covering the period between 1950 and 2003 to study the city of Muscat in Oman.

In the past few years, significant research has been carried out on the use of satellite data and Geographical Information Systems (GIS) for measuring city growth and sprawl development patterns. Herold et al. (2002, 2003) conducted a comprehensive study on urban land cover dynamics using remote sensing data for the development of spatial (landscape) metrics that were analyzed and interpreted in conjunction with results from the spatial modeling of urban growth. Remote sensing provides spatially consistent data sets that cover large areas with both high spatial detail and high temporal frequency. Dating back to 1960, remote sensing can also provide consistent historical time series data. The importance of remote sensing was emphasized as a "unique view" of the spatial and temporal dynamics of the processes in urban growth and land use change (Dewan and Yamaguchi 2009b; Herold et al. 2003; Boori and Amaro 2010). Satellite remote sensing techniques have therefore been widely used in detecting and monitoring land cover changes at various scales with useful results (e.g., Wu et al. 2013; Stefanov et al. 2001; Wilson et al. 2003; Boori et al. 2010; Muller and Zeller 2002; Weng 2002). Recently, remote sensing has been used in combination with GIS and Global Positioning Systems (GPS) to assess land cover change more effectively than by remote sensing data only (Dewan et al. 2012; Boori et al. 2015a). It has already proved useful in mapping urban areas and as data source for the analysis and modeling of urban growth and land use/land cover change (e.g., Rodriguez-Galiano and Chica-Olmo 2012; Grey et al. 2003; Herold et al. 2003; Wilson et al. 2003).

The objective of this study is to determine the spatial patterns of urban growth over time in Kuala Lumpur. Thus, the growth has been systematically mapped, monitored, and accurately assessed using satellite images in concert with conventional ground data. Mapping provides a "picture" of where growth is occurring, helps to identify the environmental and natural resources threatened by such development, and suggests the likely future directions and patterns of growth. The current study has three specific objectives: (1) investigate the growth of Kuala Lumpur City during 1989–2014 by using remote sensing and GIS; (2) analyze the existence of urban sprawl in Kuala Lumpur; and (3) evaluate current local environmental and natural resource protection and development policies.

Study area

Kuala Lumpur was chosen as the study area due to its rapid growth. Kuala Lumpur (latitude 3° 8′ N; longitude 101° 44′ E) is situated in the Federal Territory of Kuala Lumpur in the west Peninsular Malaysia (Fig. 1). The total population of Kuala Lumpur in 2014 is 1.67 million people in an area of 94 km^2. This gives the city proper a very high population density of 17,310 people per square mile or 6890 per square kilometer. Greater Kuala Lumpur or the Klang Valley is a large urban agglomeration with an estimated population of 7 million in 2014 with a population density that is nearly equal to that of the city proper (City Hall of Kuala Lumpur 2008). According to the 2010 census, major ethnic groups in Kuala Lumpur are as follows: Malays 44.2 %, Chinese 43.2 %, Indians 10.3 %, and others 1.8 %. Kuala Lumpur is an aging city as birth rates have fallen in the last decade or two. This has led to a lower percentage of young people under 15, a group that has fallen from 33 % in 1980 to under 27 % in 2000. The working age group between 15 and 59 has increased from 63 % in 1980 to 67 % in 2000 as more people move to the city for opportunities. By 2020, Kuala Lumpur is projected to have a population of 2.2 million.

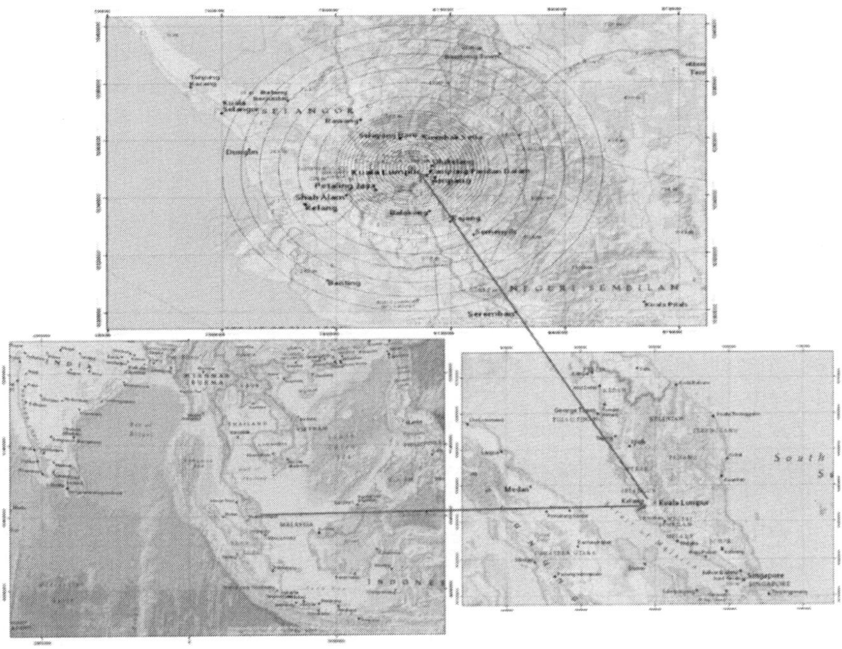

Fig. 1: Study area and multi-buffer ring zones around the city center of Kuala Lumpur

The geography of Kuala Lumpur is characterized by the huge Klang Valley. The valley is bordered by the Titiwangsa Mountains in the east, several minor ranges in the north and the south, and the Strait of Malacca in the west. Kuala Lumpur has a tropical rainforest climate, which is warm and sunny, along with abundant rainfall, especially during the northeast monsoon season from October to March.

METHODS

Data

A selection of multi-sensor and multi-temporal images was used for this study (Ding and Shi 2013; Gamba et al. 2005). The specific satellite images used were Landsat TM for 1989, Landsat ETM+ for 2001, and finally Landsat 8 for 2014, an image captured by different types of sensor. So, three different date satellite images were used with less than 10 % cloud cover. The spatial resolution is 30 m for band numbers 1 to 5 for all sensors, which is used in this research work. All images were from the USGS website, which is free of cost. All images are radiometric and geomantic corrected. As all images have the same spatial resolution for specific bands, one simply goes through maximum likelihood classification. Here, at least 10 years' time difference was used in two dates' data (Jensen and Im 2007) so that sufficient land use/cover change information can be investigated (Dewan and Yamaguchi 2008; Anderson et al. 1976). Another reason for selecting these images was their availability.

Maps processing

The images obtained as standard products were geometrically and radiometrically corrected by data providers, but, because of the different standards and references used by the various image-supplying agencies, all images were georeferenced again at the pre-processing stage. Then, all three images were geometrically corrected up to the orthorectified level. Images from different sensors have different spatial resolutions. In this study, we preferred to retain the spatial detail, original pixel size, and value of each image for specific bands. Therefore, the images were kept without changing their pixel sizes despite the possible varying accuracy level of classification with the different spatial, spectral, and radiometric resolutions. Next, at the classification stage, a uniform supervised classification was applied on the images. All of the images were classified by creating accurate polygons as training areas for introducing ideal classes for each image separately and by using the maximum likelihood classification method. To create a closer correspondence between the maps produced, the classification was done by only considering four main classes: urban built-up area, agriculture land, forest land (tree/park), and water (Butt et al. 2012; Anderson et al. 1976; Boori and Ferraro 2015). The selected main classes and their descriptions are presented in Table 1.

Table 1: Description of land use/cover classes

Land use/class	Description
Urban or built-up	Residential, commercial and services, industrial, transportation and roads, mixed urban and other urban
Agricultural and vegetation	Agriculture field, small vegetation, sand soils, landfill sites, areas of active excavation, dry salt flats
Forest or tree/parks	Pine, coniferous trees and citrus orchards, parks, grass belts
Water bodies	River, permanent open water, lakes, ponds and reservoirs, permanent and seasonal wetlands, swamps

Post-classification refinement was used to improve the accuracy of the classification. In addition, because the urban surface is heterogeneous with a complex combination of features (buildings, roads, grass, trees, soil, and water), mixed pixels and the salt-and-pepper effect are common problems when using medium spatial resolution data, such as Landsat (Byomkesh et al. 2012; Lu and Weng 2005; Boori et al. 2014c, Thinh 2003). In this step, a 3 × 3 majority filter was applied to all the classified land covers (Were et al. 2013; Lillesand and Kiefer 1999).

This research is focused on urban built-up and urban vegetation coverage measurement and their interactions. Rangeland is considered to be both non-urban vegetation and forest area. Furthermore, water bodies have very limited coverage and along with bare land cover are not discussed further in the Results and discussion and Conclusions sections. The maps produced are presented in Fig. 2.

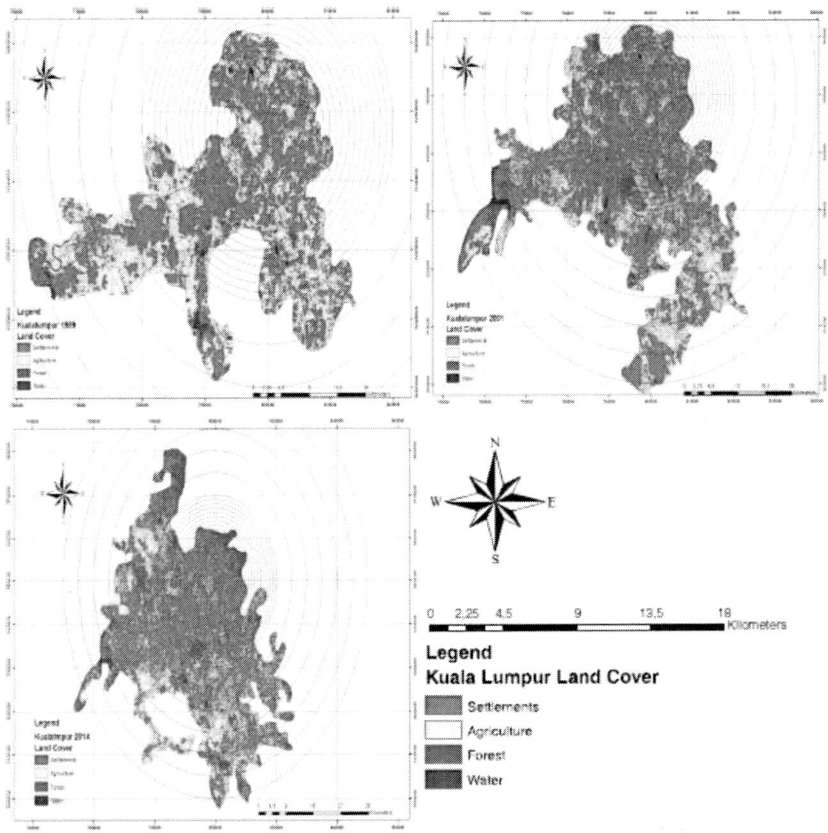

Fig. 2: Multi-buffer ring zones around the city center of Kuala Lumpur for 1989, 2001, and 2014 map

After classification, multi-buffer rings are created for every 1-km distance from 1 to 50 km from the city center to the outside. Then, the intersection with classified maps for all three dates is performed. Later on, all class areas were measured for 1- to 50-km distance and derive density according to Eq. 1:

$$\text{Urban density} = \frac{\text{Settlement area/ring}}{\text{Total ring area}} \tag{1}$$

Urban growth rate

In order to evaluate the spatial distribution of urban expansion intensity, we adapted an indicator called annual urban growth rate (AGR) for evaluating the "urbanization" speed of unit area (Xiao et al. 2006). AGR is defined as Eq. 2:

$$AGR = \frac{UA_{n+i} - UA_i}{nTA_{n+i}} \times 100\ \% \quad (2)$$

where TA_{n+i} is the total land area of the target unit to be calculated at the time point of $i + n$; UA_{n+i} and UA_i the urban area or built-up area in the target unit at times $i + n$ and i, respectively, and n is the interval of the calculating period (in years). Generally, the target calculating unit is set to the administrative district so as to link with administration or economic statistics. In this research, we preferred the geographical gridding unit since the administrative borders have been changed so frequently in this city. The maps were therefore gridded as 1 × 1 km units, and the annual urban growth rates of each unit were then calculated. Lastly, the grid-based annual urban growth rates were mapped to evaluate the spatial features of the "expansion."

RESULTS AND DISCUSSION

In this research work, more than 10 samples/training sites have been selected for each class so that classification accuracy is high. The high level of accuracy observed in this study could be the result of the coarse classification since only four classes were used. Furthermore, in all buffer rings, only the urban area was calculated in place of the whole ring. After producing complete land use maps, the total coverage of different classes was determined. Using this information, we calculated the water, forest, vegetation, and built-up area per capita for all the study years (Table 2).

Table 2: Kuala Lumpur land use/cover classes for 1989, 2001, and 2014

Land use/cover	**1989**		**2001**		**2014**	
Settlements	456.99	51.81	1098.48	60.22	1663.23	64.36
Agriculture	345.46	39.16	520.40	28.53	699.21	27.06
Forest	70.50	7.99	188.74	10.35	209.99	8.13
Water	9.18	1.04	16.54	0.91	11.68	0.45
Total	882.13	100.00	1824.16	100.00	2584.11	100.00

There are several major trends evident in the changes of land cover that are consistent over the period 1989 to 2014. The urban/built-up area increased rapidly, and there was a marked decrease in other classes. The final four categories in the land cover maps were forest, urban/built-up, agriculture, and water (Fig. 1). Some interesting patterns in the distribution of dwellings across different buffer zones can be noted in Fig. 1. As expected, the urban/built-up density in Kuala Lumpur tends to decline outward from the city center. According to the land use/cover maps produced for the urban area of Kuala Lumpur City, built-up area was expanded from 456.99 km^2 in 1989 to 1098.48 km^2 in 2001 and finally reached to 1663.23 km^2 in 2014 (Fig. 3). This figure represents 46.67 % growth in built-up area over a period of three decades. In the first decade, urban growth was 35.16 % at an average rate of 3.19 km^2/year. In the years 2001 to 2014 (1663.23 km^2), the average rate of expanse was 1.68 km^2/year and totals 21.85 % in the second decade. So, city expanse rate was reduced just half (1.51 km^2/year.) in the second decade. The urban/built-up area has grown by factor 4 over the past three decades, increasing from 51.81 % of the study area in 1989 to 60.22 % in 2001 and finally 64.36 % in 2014, while the area of agriculture land decreased from 39.16 % in 1989 to 28.53 % in 2001 and 27.06 % in 2014. The area of forest has first increased from 7.99 to 10.35 % from 1989 to 2001 and later on reduced 8.13 % in 2014. In contrast, the area of water was 1.04 % in 1989, 0.91 % in 2001, and 0.45 % in 2014 (Table 2).

City density patterns around the city center were also examined to ascertain whether different zones have represented different densities. Figure 4 shows 50 buffer ring zones from 1- to 50-km distance. It has been argued that the first five zones represent the areas that are within walking distance from the city center. High-density development has been supported by the government to encourage people to walk to train stations or other destinations such as shopping centers. This policy has in part been pursued to encourage active transport as part of a healthy lifestyle. We find that in the city center, urban density was more than 90 % for the last three decades. In 1989, it was reduced dramatically around 55 % on the distance of 5 km from the city center, less than 10 % on 15 km, and 0 % after 40 km. When comparing 1989 and 2014 on the distances of 5, 8, 15, 20, and 30 km, the city density has increased very high. If we compare only 1989 and 2001 on the distances of 8, 15, and 19 km, the city density was highly increased. From 2001 to 2014, on the distances of 4 and 30 km, density was increased. On the distance of 50 km, the city density was 0 % in the last three decades. For the whole three decades, the city density was dramatically reduced on the distance of 20 km and later on has been stable (Fig. 4).

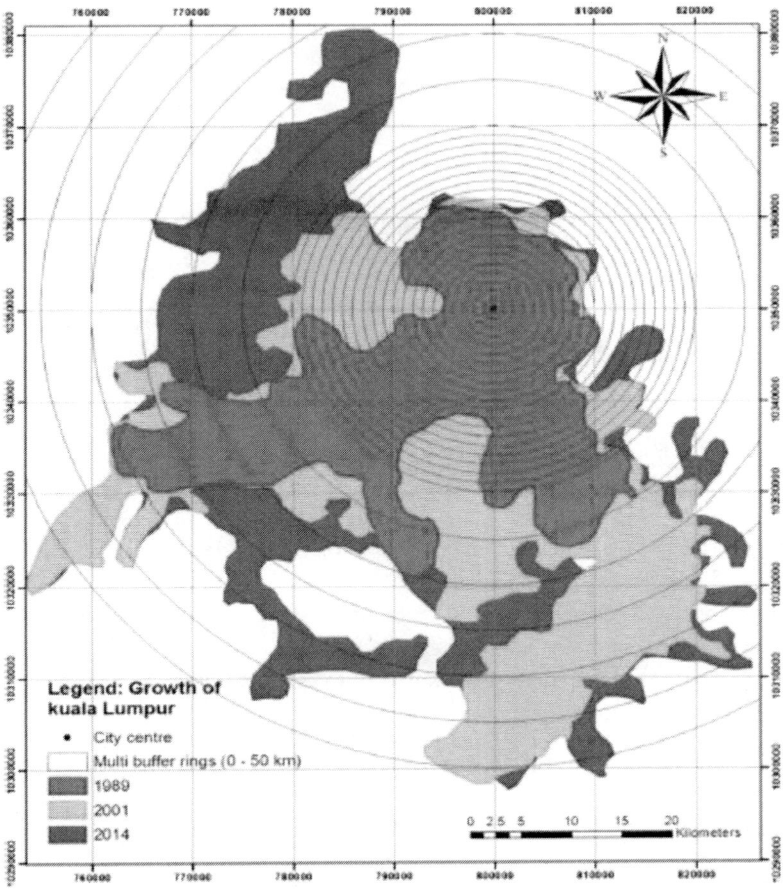

Fig. 3: Growth of Kuala Lumpur in 1989, 2001, and 2014

In 1989, the city center urban density was more than 90 %, the share of agriculture was 10 %, and forest and water was 0 % (Fig. 5). Till 4-km distance from the center, the urban/built-up area continuously reduces (55 %) and agriculture has continuously increased till 45 %. Later on, both have small variation, and from 8- to 15-km distance, both classes were reduced. Lastly, both classes were stable or vary a little, and after 40-km distance, both classes have 0 % density. The class forest has 0 % density from 1- to 4-km and from 30- to 50-km distances. It has less than 10 % density from 5- to 30-km distance. The water class has almost close to 0 % density in all distances.

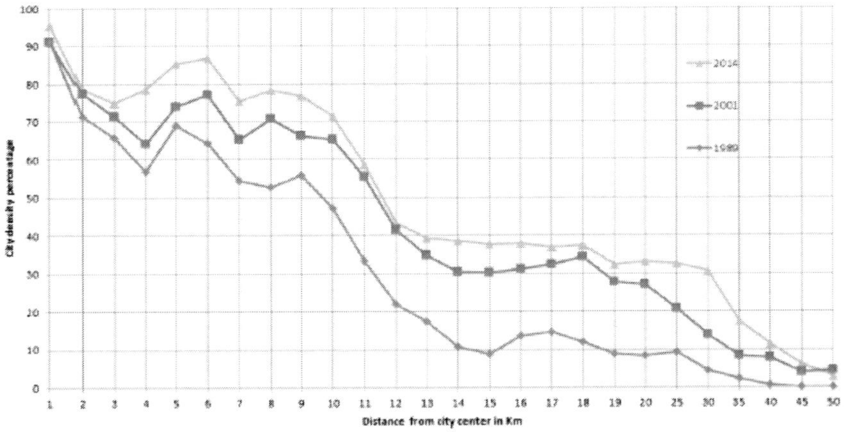

Fig. 4: Kuala Lumpur city density from 1- to 50-km distance for 1989, 2001, and 2014

In 2001, density of urban/built-up and agriculture is the same like 1989 till distance of 8 km. Later on, urbanization reduces continuously. In 2001, agriculture has 10 % density from 8- to 50-km distance. Forest has 0 % density on 1- and 50-km distance. Forest has the highest 15 % density on 9 km and 10 % density on 7-, 8-, 10-, 11-, 16-, and 17-km distance. Water class was very less on maximum distance and has 0 % density (Fig. 5).

In 2014, in the city center, urban density was more than 95 % and other classes (agriculture, forest, and water) were less than 5 %. From 1- to 3-km distance, urban/built-up areas reduce till 75 %, but agriculture increases and reaches up to 22 %, and later on reduces or varies around 10 to 15 % in all distances. Urban/built-up density again increases after 3 to 6 km and later on reduces continuously and reaches 0 % density on 50-km distance. Forest is variant on 10 % and vise versa with agriculture class. In all distances, water is close to 0 % density (Fig. 5).

The census data indicate that the main phase of population growth in Kuala Lumpur occurred between 1989 and 2001. The main causes of this growth were a high birth rate in the early years of the Islamic revolution, rural to city migration, and the merging of nearby villages with the city as it developed (Wikipedia 2014). Urban expansion and subsequent landscape changes are governed by geographical and socioeconomic factors such as population growth, policy, and economic development. In most cases, urban expansion and associated land use/cover changes resulted from a combination of these factors (Wikipedia 2014). For example, socioeconomic policy can strongly affect urban expansion and under the changes driven by urban expansion, the land use patterns of the urban outskirts are altered or adjusted in pursuit of high economic returns (orchards or croplands changing to vegetable land). Even though it is difficult to clarify the influence of these factors in these changes, their influence is examined by analyzing the relationships between developed area (including urban and residential area) and socioeconomic factors such as population, industrialization, traffic conditions, and infrastructure (Wikipedia 2014).

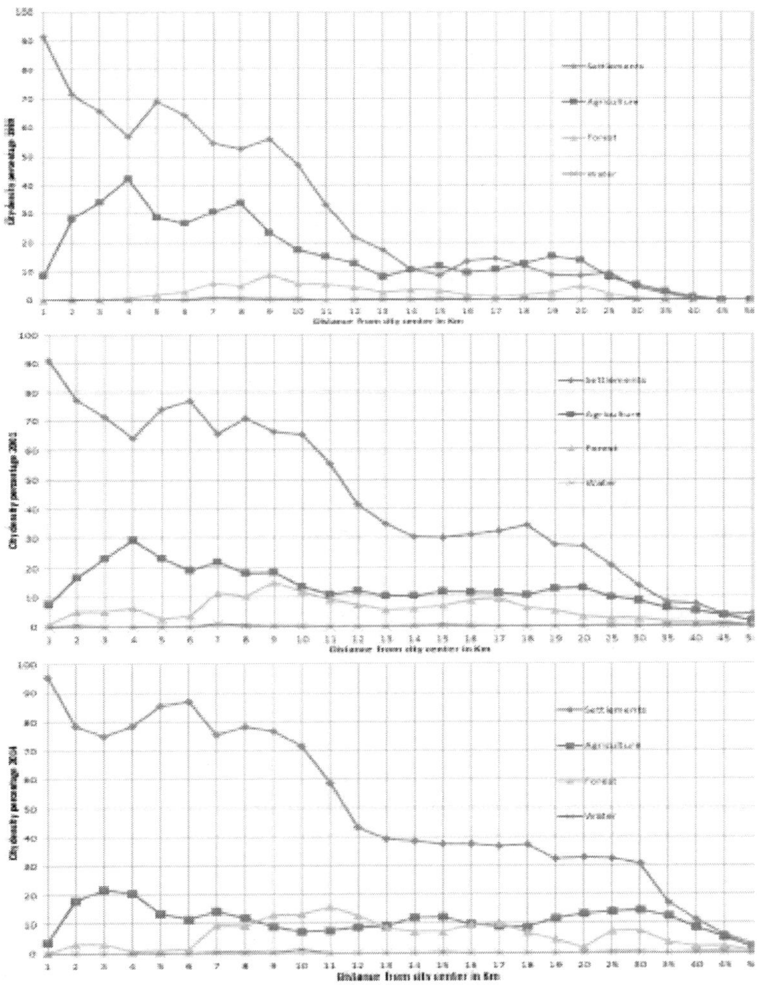

Fig. 5: Kuala Lumpur land cover class density from 1- to 50-km distance for 1989, 2001, and 2014

Asia is the most populous and arguably the most vulnerable continent in the world because of the high rate of urbanization and lack of facilities. This variability in Asia has and will continue to impact all sectors, from national and economic security to human health, food production, infrastructure, water availability, and ecosystems. Thus, many large Asian cities have become vulnerable, as more countries are unable to feed themselves and fewer countries produce exportable surpluses (Yin et al. 2011). In Asia, where most of the urban growth has concentrated in metropolitan areas, the problem of basic infrastructure and access is becoming more acute. In these urban centers, uneven distribution of incomes, the prevalence of poverty, diminishing farmlands, inefficient distribution systems, and rising expectations have all contributed to increasingly critical problems, particularly as they affect the urban poor (Dewan and Yamaguchi 2009a). A recent World Bank study maintained that as many as 360 million inhabitants of cities in developing nations suffer from chronic calorie deficits (Wu et al. 2013). Five of every six urban families in India typically spend 70 % of their income on basic needs. Even in relatively prosperous Kuala Lumpur, between 45 and 50 % of total household expenditure goes to basic needs or twice the proportion spent in the UK. Consequently, the lower-income groups in cities of Asia are often worse off nutritionally than their rural counterparts (Dewan et al. 2012).

CONCLUSIONS

Remote sensing and GIS techniques are used to monitor the dynamic phenomenon of urbanization with the help of satellite images and census data for Kuala Lumpur. In comparing 1989 and 2014 on distances of 5, 8, 15, 20, and 30 km, the city density was increased very high. In 1989 to 2001, on the distances of 8, 15, and 19 km, the city density has been highly increased. From 2001 to 2014, on the distances of 4 and 30 km, density has increased and on 50-km distance, the city density was 0 % in the last three decades. In Kuala Lumpur, urban land expansion has largely been shaped by the terrain such as mountains and lakes and social factors such as population growth, migration, and economic development. All open spaces such as gardens and vegetation cover were destroyed and converted to built-up areas; this study demonstrated that development occurred mainly in available open spaces in the city and remaining lands between the buildings. Conversion of vegetation and orchards to built-up area, however, has been a more recent phenomenon. The empirical investigation of urban residential density patterns presented in this paper offers new insights into the spatial implications of policy changes. Analysis has the potential to aid urban planners and policy practitioners in evaluations of urban consolidation or compact city policies.

COMPETING INTERESTS

The authors declare that they have no competing interests.

AUTHORS' CONTRIBUTIONS

MSB did this research work. MN provides guidance and KC help for analysis work. VV provides financial support for this research work. All authors read and approved the final manuscript.

REFERENCES

1. Al-Awadhi T (2007) Monitoring and modeling urban expansion using GIS and RS: case study from Muscat, Oman. In: Proceedings of urban remote sensing joint event, 2007. Paris, France
2. Alphan H (2003) Land use change and urbanization in Adana, Turkey. Land Degrad Dev 14(6):575–586
3. Anderson JR, Hardey E, Roach J, Witmer RE (1976) A land use and land cover classification system for use with remote sensor data. US geological survey professional paper, Washington, DC. 964., p 28
4. Barnsley MJ, Barr SJ (1996) Inferring urban land use from satellite sensor images using Kernel-based spatial reclassification. Photogramm Eng Remote Sens 62:949–958
5. Boori MS, Amaro VE (2010) Land use change detection for environmental management: using multi-temporal, satellite data in Apodi Valley of northeastern Brazil. Applied GIS 6(2):1–15
6. Boori MS, Ferraro RR (2015) Global Land Cover classification based on microwave polarization and gradient ratio (MPGR). Geo-informatics for ntelligent Transportation 71:17–37. doi:10.1007/978-3-319-11463-7-2
7. Boori MS, Vozenilek V (2014) Assessing land cover change trajectories in Olomouc, Czech Republic. Int J Environ, Ecol, Geological and Mining Engineering 8(8):540–546
8. Boori MS, Amaro VE, Vital H (2010) Coastal ecological sensitivity and risk assessment: a case study of sea level change in Apodi River (Atlantic Ocean), Northeast Brazil. Int J Environ, Earth Science and Engineering 4(11):44–53
9. Boori MS, Vozenilek V, Choudhary K (2014a) Land use/cover change and vulnerability evaluation in Olomuc, Czech Republic. ISPRS Ann Photogramm Remote Sens Spatial Inf Sci II 8:77–82. doi:10.5194/isprsannals-II-8-77-2014

10. Boori MS, Vozenilek V, Burian J (2014b) Land-cover disturbances due to tourism in Czech Republic. Advances in Intelligent Systems and Computing 303: 63–72. doi:10.1007/978-3-319-08156-4-7

11. Boori MS, Vozenilek V, Choudhary K (2014c) Land use/cover disturbances due to tourism in Jeseniky Mountain, Czech Republic: a remote sensing and GIS based approach. The Egyptian Journal of Remote Sensing and Space Sciences 17(3):01–10. doi:10.1016/j.ejrs.2014.12.002

12. Boori MS, Vozenilek V, Choudhary K (2015a) Exposer intensity, vulnerability index and landscape change assessment in Olomuc, Czech Republic. Int Arch Photogramm Remote SensSpatial Inf Sci XL-7/W3:771–776. doi:10.5194/isprsarchives-XL-7-W3-771-2015

13. Boori MS, Netzband M, Vozenilek V, Choudhary K (2015b) Urban growth in last three decades in Kuala Lumpur, Malaysia. IEEE: Urban Remote Sensing Event (JURSE), Joint, pp. 01-04. Doi:10.1109/JURSE.2015.7120536

14. Butt MJ, Waqas A, Iqbal MF, Muhammad G, Lodhi MAK. (2012) Assessment of urban sprawl of Islamabad metropolitan area using multi-sensor and multi-temporal satellite data. Arab J Sci Eng 37.1(2012):101–114

15. Byomkesh T, Nakagoshi N, Dewan AM (2012) Urbanization and green space dynamics in Greater Dhaka, Bangladesh. Landsc Ecol Eng 8(1):45–58

16. Dewan AM, Yamaguchi Y (2008) Effect of land cover change on flooding: example from Greater Dhaka, Bangladesh. Int J Geoinformatics 4(1):11–20

17. Dewan AM, Yamaguchi Y (2009a) Land use and land cover change in Greater Dhaka, Bangladesh: using remote sensing to promote sustainable urbanization. Appl Geogr 29(3):390–401

18. Dewan AM, Yamaguchi Y (2009b) Using remote sensing and GIS to detect and monitor land use and land cover change in Dhaka Metropolitan of Bangladesh during 1960–2005. Environ Monit Assess 150(1–4):237–249

19. Dewan AM, Yamaguchi Y, Rahman MZ (2012) Dynamics of land use/cover changes and the analysis of landscape fragmentation in Dhaka Metropolitan, Bangladesh. Geo J 77(3):315–330

20. Ding H, Shi W (2013) Land-use/land-cover change and its influence on surface temperature: a case study in Beijing City. Int J Remote Sens 34(15):5503–5517

21. Gamba P, Dell'Acqua F, Dasarathy BV (2005) Urban remote sensing using multiple data sets: past, present, and future. Information Fusion 6:319–326

22. Grey WMF, Luckman AJ, Holland D (2003) Mapping urban change in the UK using satellite radar interferometry. Remote Sens Environ 87:16–22

23. Herold M, Clarke KC, Scepan J (2002) Remote sensing and landscape metrics to describe structures and changes in urban landuse. Environment and Planning A 34(8):1443–1458

24. Herold M, Goldstein N, Clarke KC (2003) The spatio-temporal form of urban growth: measurement, analysis and modeling. Remote Sens Environ 86(3):286–302

25. Jensen JR, Im J (2007) Remote sensing change detection in urban environments. In: Jensen RR, Gatrell JD, McLean D (eds) Geo-spatial technologies in urban environments: policy, practice and pixels, 2nd edn. Springer-Verlag, Heidelberg, pp 7–30

26. Lillesand TM, Kiefer RW (1999) Remote sensing and image interpretation. John Wiley and Sons, New York

27. Lu Q, Weng D (2005) Urban classification using full spectral information of Landsat ETM imagery in Marion County, Indiana. Photogramm Eng Remote Sens 71(11):1275–1284

28. Malaysia (1991) Economic structural change and urbanization.

29. Muller D, Zeller M (2002) Land use dynamics in the central highlands of Vietnam: a spatial model combining village survey data with satellite imagery interpretation. Agric Econ 27:333–354

30. Rodriguez-Galiano V, Chica-Olmo M (2012) Land cover change analysis of a Mediterranean area in Spain using different sources of data: multi-seasonal Landsat images, land surface temperature, digital terrain models and texture. Appl Geogr 35(1):208–218

31. Saboori B, Sulaiman J, Mohd S (2012) Economic growth and CO_2emission in Malaysia: a cointegration analysis of the environmental Kuznets curve. Energy Policy 51:184–191

32. Salari HE, Hengsbach T, Henke C, Kopec J, Kozinski M, Sumpelmann P (2013) Detection and analysis of land use changes and growth of Hanoi using remote sensing methods and Landsat satellite images. Technische University Dortmund 2013:01–12

33. Stefanov WL, Ramsey MS, Christensen PR (2001) Monitoring urban land cover change: an expert system approach to land cover

classification of semiarid to arid urban centers. Remote Sens Environ 77:173–185

34. Thinh NX (2003) Contemporary spatial analysis and simulation of the settlement development of the Dresden city region. 17th International Conference Informatics for Environmental Protection, Cottbus, pp 253–261

35. Thinh NX, Arlt G, Heber B, Hennersdorf J, Lehmann I (2013) Pinpointing sustainable urban land-use structures with the aid of GIS and cluster analysis. Sustainability in the information society 15th symposium informatics for environmental protection, metropolis Verlag Marburg., pp 559–567

36. Weng QH (2002) Land use change analysis in the Zhujiang Delta of China using satellite remote sensing, GIS and stochastic modelling. J Environ Manage 64:273–284

37. Were KO, Dick OB, Singh BR (2013) Remotely sensing the spatial and temporal land cover changes in Eastern Mau forest reserve and Lake Nakuru drainage basin, Kenya. Appl Geogr 41:75–86

38. Wikipedia (2014) http://en.wikipedia.org/wiki/Kuala_Lumpur. Accessed date: 09/2014

39. Wilson EH, Hurd JD, Civco DL, Prisloe MP, Arnold C (2003) Development of a geospatial model to quantify, describe and map urban growth. Remote Sens Environ 86:275–285

40. Wu KY, Ye ZY, Qi, ZF, Zhang H. (2013) Impacts of land use/land cover change and socioeconomic development on regional ecosystem services: the case of fast-growing Hangzhou metropolitan area, China. Cities 31:276–284

41. Xiao J, Shen Y, Ge J, Tateishi R, Tand C, Liang Y, Huang Z (2006) Evaluation urban expansion and land use change in Shijiazhuang, China, by using GIS and remote sensing. Landsc Urban Plan 75:69–80

42. Yagoub M (2004) Monitoring of urban growth of a desert city through remote sensing: Al-Ain, UAE, between 1976 and 2000. Int J Remote Sens 25(6):1063–1076

43. Yin J, Yin Z, Zhong H, Xu S, Hu Z, Wang J, et al. (2011) Monitoring urban expansion and land use/land cover changes of Shanghai metropolitan area during the transitional economy (1979–2009) in China. Environ Monit Assess 177(1–4):609–621

8

COMMUNITY PERCEPTIONS OF CRIMINALITY: THE CASE OF THE MALTESE WALLED CITY OF BORMLA

JosAnn Cutajar[1,*], Saviour Formosa[2] and Trevor Calafato [2]

[1] Department of Sociology, Faculty of Arts, University of Malta, Room 253, Old Humanities Building (OH), University of Malta, Msida MSD 2080, Malta

[2] Department of Criminology, Faculty for Social Wellbeing, University of Malta, Humanities A (Laws, Theology, Criminology), Msida MSD 2080, Malta

ABSTRACT

This paper explores the situational factors that explain why crime in Bormla occurs in certain areas and not others. Bormla, one of the oldest cities in Malta, is the fifth crime hotspot on the Maltese Islands. The objective of this paper is to find out whether socio-demographic aspects and/or other contextual aspects are linked with the rate and type of crime that takes place within this city. A multi-method approach was adopted for the purposes of this paper. Statistics issued by the Malta police were analysed to find out which criminal offences occur there, and in which location. This analysis took place with the help of Geographic Information System (GIS) software. Qualitative and quantitative data collected through a needs assessment exercise conducted with a sample of Bormla respondents in 2009 and 2010 was also analysed to find out whether residents concurred with this official picture of what type of crime takes place there, and where it occurs. Residents were also asked whether they felt safe living in this crime hotspot, when and where they felt safe, and why.

Keywords: criminality, social cohesion, geographical information systems, needs assessment, perceptions of crime, Malta

1. INTRODUCTION

This paper adopts a multi-method approach to find out which situational factors are linked with crime in Bormla. Official statistics of crime were analysed using Geographic Information System (GIS) software. The results were compared and contrasted with data derived from a "needs assessment" study conducted in Bormla.

The needs assessment study [1] utilised a qualitative and quantitative approach to collect data. The ethnographic part consisted of interviews with service providers working with people from the area and participant observation conducted over a period of seven years. The survey was conducted with 317 respondents, representing 7% of the population aged 18 and over registered as living in Bormla in 2009. The respondents in the sample used in this survey were slightly older than the national average, while 55% were female, a rate higher than the national average. The majority of these respondents had lived in Bormla all their life (50%) or had been living there for more than ten years (41%). The majority had a secondary level of education or lower, and tended to live in rented apartments. Their level of education, coupled with the type of accommodation and tenancy, indicate that the participants were more likely to belong to a lower socio-economic category.

2. BORMLA'S SOCIO-DEMOGRAPHIC CHARACTERISTICS

This paper is based on the premise that crime and victimisation are rooted in the social and physical characteristics of communities [2], an approach adopted by environmental criminology. This approach was adopted in this paper as a means to find out which situational factors are linked with the particular pattern of crime found in Bormla. This section will begin by presenting some of the socio-demographic characteristics of the context under scrutiny. The interest in Bormla arises from the fact that this town is often associated by the Maltese general public with crime since a substantial number of convicted persons come from this area, as will be delineated below.

Bormla exhibits high population density, high dwelling density and the highest offender rate in the Maltese Islands [3]. Prior to World War II, it was Malta's industrial hub, as the British naval shipyards were located in the area. Bormla started experiencing high population turnover and a shrinkage in its population during and after World War II. Bormla suffered intensive bombing during this war because the shipyards were the main target of aerial bombing. This town is physically segregated from the rest of Malta by a system of bastions built during the reign of the Order of St. John which ruled over Malta from the 16th to the 18th century.

The resultant destruction caused by the war forced a number of residents of this area to seek shelter in safer areas in the Maltese Islands. Those who moved out of the area tended to be families with the material means to find alternative shelter elsewhere. A good portion of these wartime refugees did not return after the war [4]. The war led to a structural change in the demographic and social composition of the area. Boswell maintains that the "post-war repopulation of the Cottonera is said to have introduced a poorer and socially more depressed working-class population than it had before its elite moved out" [4].

The post-war era saw massive reconstruction programmes taking place in the area. A number of social housing projects in Bormla were built by the British in areas that had suffered war damage. Other social housing projects were built after independence (Malta became independent in 1964). The majority of social housing units in the Maltese Islands are concentrated in Valletta and Bormla, according to statistics derived from the Maltese Lands Department. In fact, 8.8% of the social housing units in the Maltese Islands are to be found there, according to the Maltese Department of Lands [5] (the Maltese social housing agency). Unfortunately, the architecture of these social housing projects clash with the medieval and baroque architecture prevalent in the area.

The propensity for destroying old houses and replacing them with blocks of flats to accommodate families with social problems means that Bormla was, in 2011, still an area with a high population density at 5,784 persons per km^2 as compared to 7,125 in 1995 [6]. It is also an area which suffers from a lack of open spaces. There are also a number of dilapidated and abandoned buildings which residents avoid since drug-related activities tend to take place in such areas and serve as "no-go" areas for the locals. Some of the mediaeval and baroque historic buildings found here are inhabited by squatters who have no inclination (nor sometimes the means) to renovate both the place and its surroundings [7].

Old houses within the core of the city are rented out at cheap rates and it is not unusual for a house to be sub-divided into separate units and rented to different households. Since the rent for this sub-standard housing tends to be low, materially deprived groups tend to be attracted to the area [8]. Long-term residents regard the people who move into Bormla as "outsiders" [9] and a "scourge" since, according to them, their behaviour gives the place a bad name. The "locals" blame policy makers for building flats in an area which was already overpopulated and then "dumping" problem families there. Locals and service providers blamed some local councillors from neighbouring towns for this state of affairs. They also cited a list of families from these neighbouring towns which had been given alternative social housing in Bormla in order to "clean up" their own town. When this issue was raised in an interview with one of the councillors in question, this person stated that these families were persuaded to leave the area, but did not go into detail on the type of persuasion used.

"Outsiders" provided with social housing in Bormla might not feel comfortable living there. Some mentioned the fact that they were now cut off from the social networks they had made in their town or village of origin, social networks which are essential for survival for people living at or below the official poverty line. Newcomers also find it difficult to integrate into this close-knit community, where most people are related to each other, and practically everybody knows each other's family history and circumstances.

To make matters worse, people living in the area have been badly hit by the economic restructuring which took place in the Maltese Islands in the first decade of the 21st century. A good portion of Bormla residents used to work in the dockyards and shipbuilding industries [10]. Their numbers decreased drastically when these were privatised in 2010. Another segment of the population worked within the manufacturing industry. The re-location and down-sizing of a number of factories in the first decade of the millennium [7,11], left a number of people from the area without employment, or precariously employed.

The unemployment rate in the Southern Harbour area in the first decade of the millennium was comparatively high [12]. It therefore came as no surprise when the Southern Harbour district (Bormla is one of the localities found in this district) ranked second worst in the 2007 Survey on Income and Living Conditions (SILC) [13] and in subsequent ones. In the SILC survey, 17% of the residents living in the Southern Harbour district were found to be living at or below the poverty line. In this survey, the Gozo and Comino district was found to have the highest percentage of people living at or below the poverty line (18% according to NSO [14]). The Gozo and Comino district, however, did not have as high a crime rate as Bormla.

Ameen [15] states that the number of people dependent on state handouts is higher in Bormla than in other areas of Malta. This high rate of dependence on social welfare is due to the high rate of unemployment found in the area, as noted above. The high unemployment rate is also linked with the comparatively higher illiteracy rate. Data elicited from the 2005 census demonstrates that the percentage of illiterate Bormla people was 14.9%, more than double the national average (7.2% of the population, according to Galea [16]). Although attempts have been made to ensure a better academic future for the young people of this locality [17], educational attainment among the young of Bormla might not be reversed in the future since this is a locality which suffers from a high rate of school absenteeism when compared to other areas of Malta [18].

Formosa [19] also notes that Bormla hosted the highest offender rates between 1980 and 1999. Before this period, Valletta was the city with the highest rate of offenders (Table 1). This concentration of registered offenders in Bormla might have been caused by social housing placements as well as the availability of cheap accommodation which might have attracted those without enough material means to find rented accommodation in the area.

Table 1. Offender Residence Changes: 1950–1999.

Locality	Offender % 1950 to 1959	Offender % 1960 to 1969	Offender % 1970 to 1979	Offender % 1980 to 1989	Offender % 1990 to 1999
Bormla	6.3	5.4	9.3	11.1	14.2
Valletta	10	11.8	17.4	9	12.7

Source: Formosa [3].

Figure 1 shows that the majority of offenders living in Bormla tended to live in poverty hotspots, namely, those poverty hotspots which registered 3.5 times the national rate [3]. Ex-offenders are more likely to face poverty because, as Camilleri [20] points out, they are less likely to find work once they come out of prison. The spatial correlation resultant from the analysis of the two datasets highlights the fact that the areas where poverty is concentrated also host the offender residences. In addition, when offenders leave prison, those who do not return to their source town migrate to poor areas in other towns, mainly Valletta.

This concentration of ex-offenders in the area may be high also because it has been attested that materially deprived people are more likely to be arrested, to be charged by the police, to be denied bail, to appear in court without adequate legal representation, and therefore more likely to be convicted [4]. One should, however, note that material deprivation as such does not push people into committing crime or other forms of antisocial behaviour.

Sampson, Raudenbush and Felton [21] maintain that higher crime rates in socially deprived areas might derive from a lack of collective efficacy. Constant population turnover, especially in certain social housing estates, can lead to a lack of collective efficacy. This lack of collective efficacy is, however, prevalent in certain social housing estates and not in others. The data analysed in this paper demonstrates that, in Bormla, crime is linked with poverty, lack of collective efficacy, and of windows of opportunity.

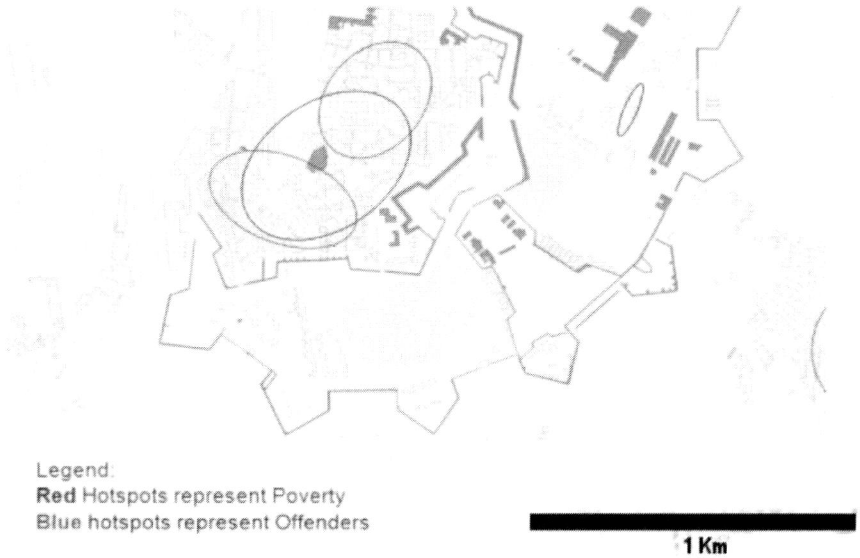

Figure 1. Crime Maps—Overlaying Offender and Poverty Hotspots.

3. OFFICIAL CRIME RATE

In this section, data provided by the CMRU (Communications and Media Relations Unit) of the Malta Police Force will be analysed to find out what the official crime rate in Bormla was in 2009 when compared to the rest of Malta, and what types of crime occurred there. The statistics provided by the CMRU depend on reports made to the police. One should point out that not all reports are registered when antisocial behaviour comes to police attention.

When one compares Bormla's crime rate with that registered for the rest of Malta, one finds that the crime rate, as registered by the CMRU in 2009, was higher than the average national figure for crime, as Table 2 attests. One, however, should point out that the first four primary crime hotspots in the

Maltese Islands in 2009 were Paceville (St. Julian's), followed by St. Paul's Bay, Sliema, Valletta and then Bormla [19].

Table 2. Official crime rate in the Maltese Islands and Bormla in 2009.

National Population	395,000
Crimes	11,953
Crimes per 1000 persons	**30.26**
Bormla Population	5,600
Crimes	187
Crimes per 1000 persons	**33.39**

Source: Formosa [22].

The main types of crime recorded in Bormla between 2004 and 2009 were mainly theft, followed by damage, bodily harm and attempted offences. When the crimes committed were collapsed, it was evident that crime against the person was higher than the national average, while crime against property was lower (Table 3).

Table 3. Type of crime committed in Bormla in 2009 compared to national figures.

Type of Crime	Bormla	National figures
Crime *vs.* property	64.7%	76.5%
Crime *vs.* person	26.2%	17.7%
Crime *vs.* general public	0%	0.8%
Attempted offences	1%	5.0%

Source: adaptation of data provided by CMRU.

3.1. Areas in Bormla Where These crimes Are Committed

As Figure 2 demonstrates, the registered crime rate in Bormla was highest in the areas where the poverty rate was high [22,23]. In terms of offence analysis, central Bormla is a relatively high concentration area of crime, as depicted in Figure 2. This analysis also shows that in comparison to other areas in the Maltese Islands, Bormla has a hotspot that covers the entire urban area [3].

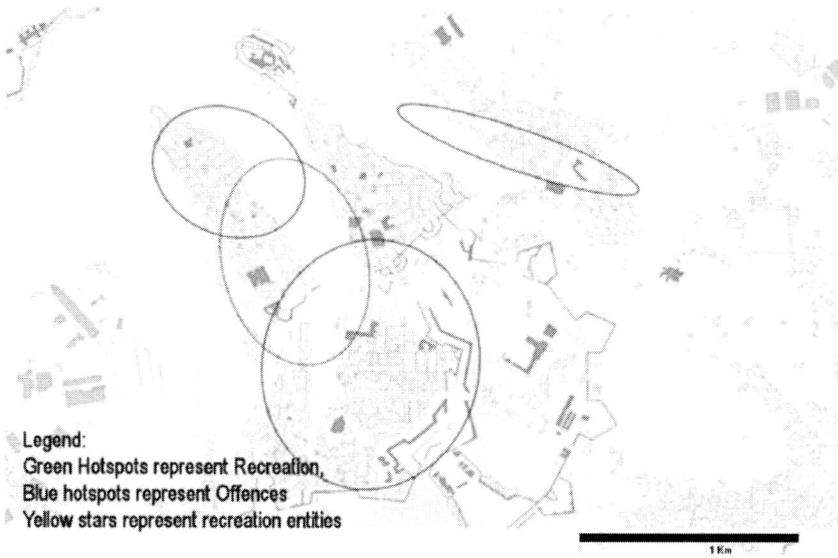

Figure 2. Crime Maps—Overlaying offence and Retail/recreation Hotspots.

Formosa [3] combined the NNH analysis with Moran's I Spatial Autocorrelation for poverty rate and offender density. The results show that there is a clustering of both poverty and offenders, at a Moran's I of 0.028094 for poverty which is higher than that for offenders at 0.009482. A spatial analysis using 1NNH hotspots at 1 standard deviation indicates that 95.2 percent (37) of the forty 1990s offender hotspots are located within or intersect with poverty areas [3].

Offence hotspots in Bormla are spread across the urban area but concentrations exist, located in an arc around the church, with two distinct clusters: one around the southeastern bastions, and another closer to the Bormla primary school and the Verdala social housing estate (Figure 3). The school is located in a non-residential area, and this might be the reason why crime here prevails. The Verdala social housing project is also located on the outskirts of Bormla, but one cannot say that the area is uninhabited. Verdala is a fort built by the British during the 19th century to protect them from a "native" insurrection. After the war, it was turned into a temporary social housing estate until more appropriate housing quarters could be provided. Unfortunately, this never took place. This area is somewhat cut off from the rest of Bormla since it is a fort. A small number of families have been living there for a number of years, but the majority do not stay long because this area is associated with antisocial behaviour. Although the Verdala housing estate has been recently renovated, the fort was not built to provide family accommodation.

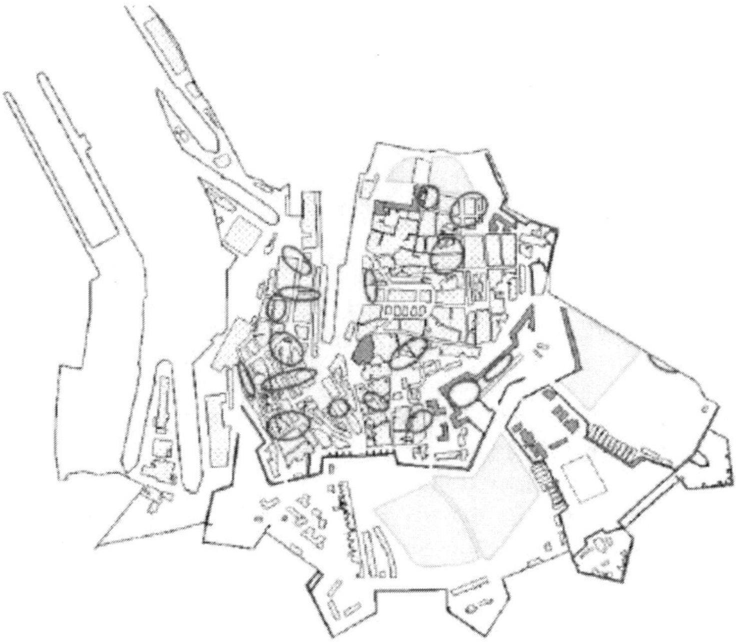

Legend:
Red Hotspots represent Offence Hotspots

1 Km

Source: Formosa [3].

Figure 3. Crime Maps—Bormla Offence Areas: 1NNH.

It is evident to those who are familiar with the geographical and social layout of Bormla, that criminal activity in 2009 took place in areas where there was a high concentration of residents living in social housing units or in privately rented accommodation (for example, an area called *Fuq San Pawl*). This, therefore, leads one to conclude that a lower rate of collective efficacy exists in areas where residents might not have had the time or capacity to form social networks with other people living in the area since they are constantly changing abode.

A high prevalence of crime also took place in commercial areas (e.g., Bormla Wharf and Gavino Gulia Square) and in less frequented public spaces (examples include Santa Margerita Square and Three Cities Road). Crime rates were also high in areas where there was a prevalence of neglected and abandoned houses such as Alessandra Street and Matty Grima Street, areas where a number of drug-related crimes took place, according to respondents who participated in the needs assessment study. Crime rates were lower in residential areas where there was a prevalence of home owners, and/or residents who had been living there for generations.

3.2. The Community's Perception of Crime

Statistics give one interpretation of crime. It is also imperative to look at the community's perception of crime since their experience of antisocial behaviour might be different from that which emerges from statistical analysis. For the purposes of this study, data from the Bormla needs assessment study will be discussed here. This section will focus on what residents feel about living in one of the crime hotspots in Malta.

When a sample of Bormla residents were asked in the needs assessment survey what types of antisocial behaviour took place in the areas where they lived, the four prevalent criminal and antisocial behaviours mentioned were drug-related crimes, unsupervised youth, followed by theft from vehicles, and theft of vehicles (Figure 4). Although three of these crimes feature in the official statistics, it was surprising to note that residents were quite concerned about the neglect of children and young people living in the area, which they felt was almost as prevalent as drug-related crime. Their concern emanated from the fact that unsupervised youth were more likely to resort to antisocial behaviour. In fact, the local authorities were at one time very concerned about the fact that some of these children and adolescents were being used by drug pushers to deliver drugs. The residents were also afraid that the prevalence of drugs in the area might mean that young people living there might have more opportunity to dabble with drugs. Substance abuse not only has a deleterious effect on the person who abuses, but can also lead to crime.

The residents' perception of crime in this survey does not concur with the crime statistics provided by CMRU, which are found in Table 3. It is quite clear that the authorities were more likely to record and act on reports that concerned crime against property than those that concerned crime against persons. This issue was raised by a police officer himself who, when asked what type of antisocial behaviour prevailed in the area, mentioned domestic violence in the same breath as drug-related crime. Although the officer in question said that the police were constantly called to deal with incidents involving domestic violence, few of these incidents were officially recorded. This low official record for domestic violence might emanate from the fact that where this type of crime is concerned, police personnel in Bormla try to reconcile the two parties and/or involve other agencies.

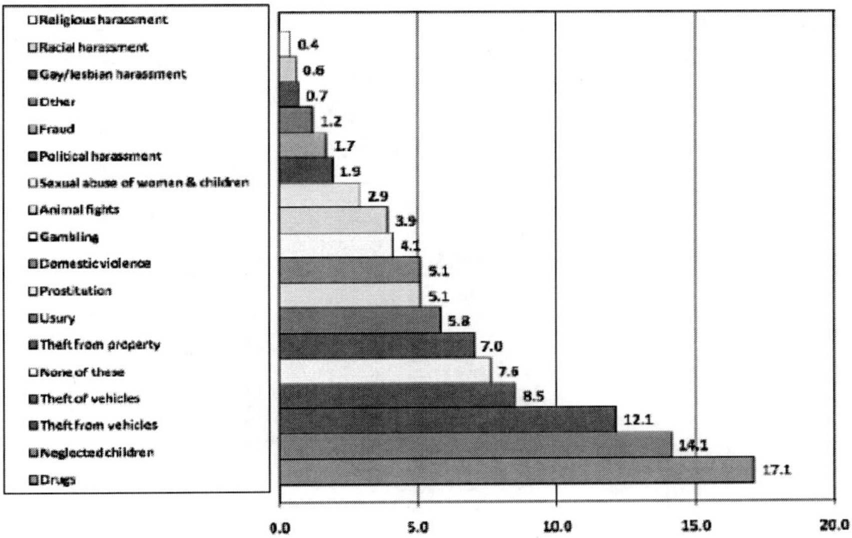

Figure 4. Criminal and antisocial behaviour in neighbourhood.

In another section of the needs assessment survey, respondents were asked whether they or a family member had been a victim of crime in the 12 months prior to the survey. Around 17% of the respondents said that they or a family member had been a victim of crime or antisocial behaviour. When one compares the victimization rate found in Table 2 (33.39 crimes per 1,000 population) with this result, one realizes that the respondents' victimization rate was significantly higher.

Respondents were also asked whether they had contacted the police in the 12 months before the survey was conducted, and 31.5% of respondents replied in the affirmative (Figure 5). This result shows that this sample of residents contacted the authorities even when the crime or the antisocial behaviour did not concern them directly. One could therefore say that they were looking out for the interests of all those living in their neighbourhood. They were, in a way, acting as an informal neighbourhood watch.

A third of the respondents who said that they had contacted the police 12 months prior to the survey said that they were not satisfied with how the issue was dealt with. These respondents felt that the police had not done enough to settle the case, or did not show any interest in tackling the case when the incident was reported. These respondents felt that the authorities needed to do more to tackle the antisocial behaviours mentioned above. Some of the respondents answered that the police needed to make their presence felt in the area by conducting more foot and car patrols. As some of the respondents pointed out, the police only turned up when something went extremely wrong.

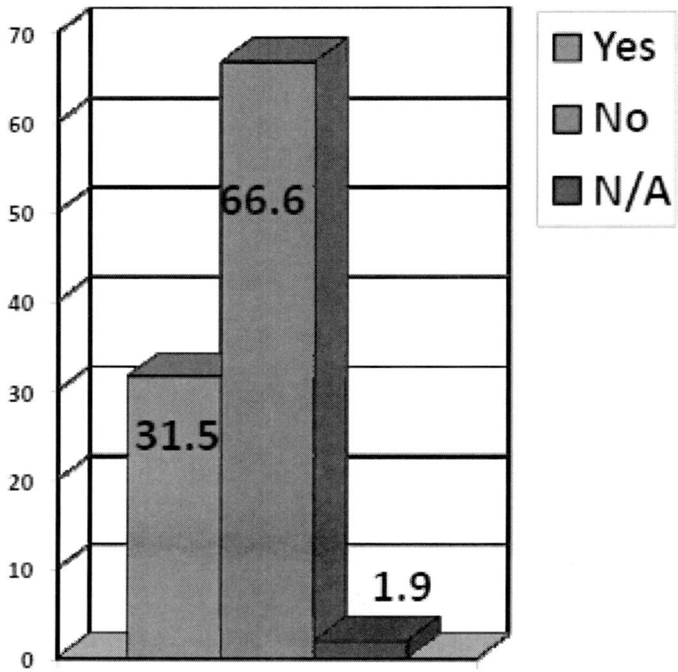

Figure 5. Did you contact the police in the last 12 months?

3.3. Perception of Safety

Although official statistics place Bormla as a crime hotspot, residents felt relatively safe living there. In fact, according to this survey, around 66.8% of respondents indicated that they felt safe living in Bormla (Figure 6). They felt safe in their own home by day, although this sense of security dropped at night. They also felt safe in the area where they lived, although this feeling of safety differed by area, marital status, gender and age. As Figure 6 demonstrates, respondents felt safer during the day. They felt safer in inhabited areas, and less safe in uninhabited ones. Female respondents were more likely to state that they felt safe in their neighbourhood and home by day than their male counterparts. They were, however, more likely to state that they did not feel safe at night.

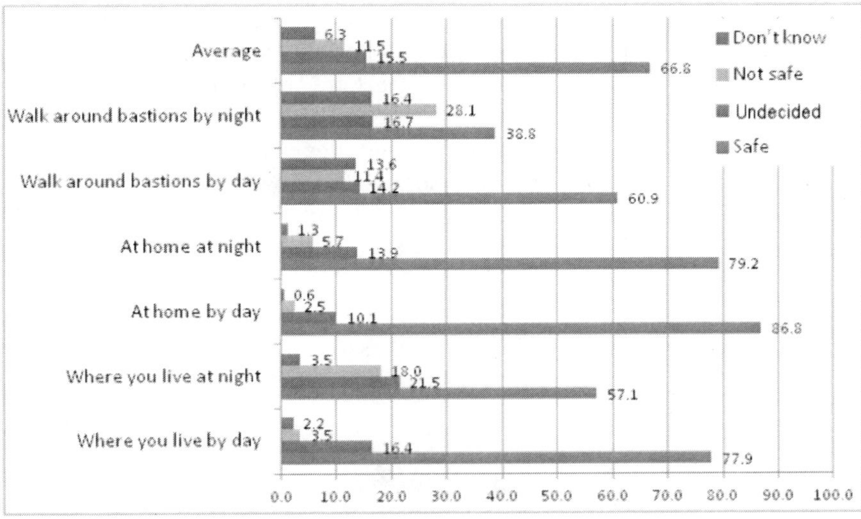

Figure 6. Perception of safety by time.

The perception of safety also differed by area. Those living in the area of Santa Margerita (which includes the streets of St. George, St. Margaret, Narrow, and Windmill), as well as those living near St. Helen's Gate (covering the area from the parish church up to St. Helen's Gate) felt slightly less safe in their neighbourhood during the day than respondents living in other areas (Figure 7). The areas mentioned here include tracts of non-residential space. Residents living in the St. Theresa area (incorporating Matty Grima, Nelson, St. Lazarus, Alexander, and St. Theresa Streets), Cottonera (including St. John t'Ghuxa, St. Nicholas, and Polverista) and the St. Helen area felt more insecure at night. Respondents living in the St. Helen area experienced a sense of insecurity both during the day and even more during the night. Those living in the St. Theresa residential area felt secure during the day, but then drug-related activities at night prompted residents to seek shelter inside their own homes. Although the Cottonera area was not linked by the police with criminal activities, residents stated otherwise. Some of the streets and lanes in the St. Theresa, St. Margaret, and Cottonera areas were considered as "no-go" areas by respondents for these reasons.

With regards to age, respondents aged been 37 to 60 were more likely to mention that they felt less safe in their neighbourhood and at home. This was especially true of residents who lived on their own. Older residents who felt safe were those who sought shelter with relatives or at a home for the elderly.

When respondents were asked to mention the areas in Bormla where they did not feel safe (the areas in question are marked in red in Figure 7), they were more likely to mention the area within and surrounding the Verdala Housing Estate. The ring road around Bormla, starting from Polverista, passing by Cottonera Gardens up to the Convent of St. Theresa was also listed as an unsafe area. Few people live there. A number of bus stops are located on this route. Respondents who depended on public transport underlined the fact that they did not feel safe waiting on the bus stops located on this route, especially at night and in the early hours of the morning.

Figure 7. Areas where residents do not feel safe.

There were also some residential areas which were not considered safe. These were the areas where a number of bars are found, particularly the one near St. Helen's Gate. Other bars are located close to the local police station. Some of the respondents living near these bars remarked that they did not feel safe when drug-related transactions took place in this area. A number of respondents also mentioned the fact that some of the patrons who frequented these places made a lot of noise, especially at night. Another group of respondents stated that some dilapidated and abandoned houses were often used by drug addicts to inject and/or "prepare" drugs.

Respondents also mentioned *Fuq San Pawl*, Matty Grima and Lazarus streets. The police were aware that drug-related transactions were taking place in these areas. They also mentioned some of the difficulties they met when it came to prosecute some of the drug dealers they arrested. Some mentioned that charges were dropped when "powerful" individuals intervened.

4. DISCUSSION

4.1. Crime Rate and Crime Hotspots

When one compares the crime hotspots mentioned by residents with those which emerged from the GIS maps (Figure 7), one notes some correlations. Some of the areas which featured in the GIS maps were mentioned. As one notes when comparing the crime hotspots (marked in red in Figure 7) with the respondents' feelings of lack of safety (marked in blue in Figure 7), the feeling of lack of safety was not always linked with the level of criminal activity taking place in the area as registered by the police. This feeling of safety or lack of safety was also affected by the time of day, unilateral or multilateral use of space, age, gender and/or marital status of respondent. It was also related to the residents' knowledge of what went on in the area.

Respondents felt that the crime rate in Bormla was higher than the rate recorded in official statistics. This issue was raised by residents in a meeting with the Police Commissioner which took place in April 2010. Residents felt that the police were not doing enough to stem the tide of illegal and antisocial behaviour prevalent in Bormla. A number of them complained that the police failed to act when they reported illegal or antisocial behaviour taking place in certain areas of Bormla.

One should also note that while respondents and the police felt that crime against the person (neglect of children coupled with domestic violence) and drug-related crime were high in Bormla, official statistics denote otherwise. Official crime statistics listed crime against property as being the most prevalent crime in Bormla, while respondents and the police mentioned crime against the person.

4.2. Who Commits Crime?

Bormla residents tend to attribute crime to the presence of "outsiders," namely, people who have ended up living in Bormla either because they were sent there by the social housing authorities or because they could not afford to rent accommodation elsewhere. They believe that a stop to "social dumping" by the local authorities would help to minimise this crime rate. Statistics on the prevalence of ex-offenders living in the area cited by Formosa [19] might give some credence to this perception. Formosa states, that prior to the 1980s, ex-offenders were concentrated in Valletta. These were relocated to Bormla during the period when the Valletta slum clearing project was enacted.

4.3. Where Does Crime Take Place?

As the data discussed above demonstrates, crime was more likely to occur in public spaces with single or low use by the community. These areas included commercial or leisure areas and other single-purpose uses of public space such as schools and circular roads. Within multi-purpose areas, crime rates were perceived to be higher in areas with a higher rate of abandoned houses, and near blocks in social housing estates. Areas with a preponderance of social housing and rented accommodation were more likely to have a high population turnover, and this might explain the lower community capacity in this area. As the data mentioned above demonstrates, there is a positive correlation between population, type of tenure, type of housing available and the crime rate.

4.4. Reasons for the High Crime Rate

Residents felt that the crime rate was high because the police rarely made their presence felt in the locality. The respondents lamented the fact that the police were rarely seen patrolling the area and that they only turned up when something went extremely wrong. They felt that they were inadequately protected by the police because as a community, as they reported, "We don't count."

Stiles *et al.* [24] correlate higher rates of violence, property crime and drug use with lower feelings of self-worth. Box [25], however, maintains that victims who come from stigmatized groups find it harder to persuade others that they are victims. This seems to be the case here. Croall [26] adds that the processes of victimization are socially, economically and culturally situated, as is true for Bormla. Low-status groups are more likely to be placed at the lower end of the victimisation hierarchy although they are more likely to be arrested, as statistics demonstrate. Data derived from the needs assessment study demonstrates that residents felt that they were less likely to be perceived as victims, and that their demands for protection were not being taken into consideration. This made them feel neglected by the authorities concerned. Shiell and Zhang [27] attest that this feeling of powerlessness is prevalent in depressed, socially excluded areas.

Agnew [28] adds that neighbourhoods with a concentration of socially disadvantaged people living in bad social conditions tend to have a concentration of angry and frustrated residents who might express this frustration by becoming physically violent. This statement might explain why the crime rate against the person in Bormla was higher than the national average. It might also explain why the crime rate in Bormla was higher when compared to other districts and/or areas having, statistically speaking, high concentrations of people living at or below the poverty line. Baron *et al.* [29] note that violence tends to prevail in urban settings where people are exposed to a combination of stresses caused by economic deprivation, urban living, and perceived discrimination, which might push some to resort to violence in interpersonal relations, a phenomenon referred to as the "culture of exasperation." Kubrin and Weitz [30] add that neighbourhoods with higher rates of dilapidation and deviance are more likely to be stigmatized.

The crime rate seemed to be higher in poorly supervised areas, namely, areas with a high population turnover, areas with a concentration of dilapidated buildings, and non-residential areas. Respondents were also more likely to link crime with night time.

4.5. Perception of Safety

Although residents felt that the crime rate in Bormla was high in comparison to other areas, they felt safe. Perception of safety is linked to place, as demonstrated above. This perception of safety was influenced by age, marital status, gender and time. Certain areas were perceived as highly unsafe by some of the residents, especially areas where public use of the area was uni-directional. Only seven out of the 317 respondents felt unsafe "*kull imkien*" (everywhere) in Bormla, and this fear emanated "*minhabba d-drogi*" (because of drug-related crime).

In spite of the high population turnover in Bormla [7], the data demonstrates that there was still a high degree of collective efficacy. This level of collective efficacy was more prevalent in areas where residents had been living for a number of years. Sampson *et al.* [21] refer to collective efficacy as the neighbours' effort to look out for each other. This level of safety was attained when residents looked out for their family's and neighbours' interests. Kaplan *et al.* [31] maintain that there is a correlation between collective efficacy, social cohesion, and income inequality.

5. CONCLUSIONS

This article looked at the rate and type of crime that took place in Bormla in the period between 2009 and 2010. Respondents felt that the rate and the type of crime cited by the authorities did not concur with their own experience of antisocial behaviour. At the same time, a good number of crime hotspots underpinned by administrative data were also identified by respondents in the needs assessment exercise.

The crime patterns which emerged in the analysis of official statistics were not always in keeping with how community members experienced crime in this location. While both law enforcers and a sample of community members agreed that crime was more likely to take place in spaces with unilateral usage, areas with a concentration of uninhabited and derelict buildings, and areas experiencing constant population turnover, they were less likely to agree on the extent and type of crime that took place in these areas. It was also evident that community members were concerned about crime against the person, while police authorities were more likely to act on reports of crime against property.

A sample of the Bormla community felt that the high crime rate in this locality was mainly caused by social dumping, coupled with a low police response rate. Statistics demonstrate that there was a high concentration of ex-offenders living within the locality, specifically in the so-called "poverty hotspots" indicated by the maps, which were incidentally also the areas with a high concentration of crime. This means that, to some extent, official statistics give some credence to the community's perception that "outsiders" were the problem, although not all crime could be laid at their door.

REFERENCES AND NOTES

1. JosAnn Cutajar. *Bormla: A Community Needs Assessment Study*. Qormi: Outlook Co-op, 2013. Eamonn Carrabine, Paul Iganski, Maggy Lee, Ken Plummer, and Nigel South. *Criminology. A Sociological Introduction*. New York: Routledge, 2004.
2. Saviour Formosa. "Spatial Analysis of Temporal Criminality Evolution: An Environmental Criminology Study of Crime in the Maltese Islands." Unpublished PhD Thesis, University of Huddersfield, UK, 2007.

3. David M. Boswell. The social prestige of residential areas. In *Maltese Society: A Sociological Inquiry*. Edited by Ronald G. Sultana, and Godfrey Baldacchino. Msida: Mireva, 1994, pp. 133–62.

4. "Government of Malta. Land and Public Registry. Lands Department." Available online: http://www.landpubreg.gov.mt/Page.aspx?pageid=772&lid=1 (accessed on 20 November 2010).

5. National Statistics Office. *Census of Population and Housing 2011: Preliminary Report*. Malta: NSO, 2012.

6. JosAnn Cutajar, and John Vella. "The role of cultural events in promoting community cohesion: the case of two socially deprived areas in the Maltese Islands." In Paper presented in *Attractions and Events as Catalysts for Regeneration and Change*, TTRI/CTCC Conference, University of Nottingham, Nottingham, UK, 24–25 September 2008.

7. Christopher S. Dunn. Crime Area Research. In *Crime: A Spatial Perspective*. Edited by Georges-Daniel E. Abeyie, and Keith D. Harries. New York: Columbia University Press, 1980.

8. Times of Malta. "Trial put off as defendant has no lawyer." *Times of Malta*, 23 January 2013. http://www.timesofmalta.com/articles/view/20130122/local/Trial-put-off-as-defendant-has-no-lawyer.454286.

9. Steven E. Barkan. *Criminology: A Sociological Understanding*, 4th ed. Upper Saddle River: Prentice Hall, 2006.

10. JosAnn Cutajar. Social Inequality. In *Social Transitions in Maltese Society*. Edited by Cutajar JosAnn, and George Cassar. San Gwann: Book Distributors Limited, 2009, pp. 225–58.

11. Planning Authority. *Development Brief for the Regeneration of the Cottonera Waterfront*, 1997. Available online: http://www.mepa.org.mt/file.aspx?f=899 (accessed on 1 July 2012).

12. National Statistics Office. *Census of Population and Housing 2005. Volume 1: Population*. Valletta: National Statistics Office, 2007.

13. National Statistics Office. *Survey on Income and Living Conditions 2007*. Valletta: National Statistics Office, 2009.

14. National Statistics Office. *Survey on Income and Living Conditions 2010*. Valletta: National Statistics Office, 2012.

15. Per-Olof H. Wikström, and Rolf Loeber. "Do disadvantaged neighbourhoods cause well-adjusted children to become adolescent delinquents? A study of male juvenile serious offending, individual

risk and protective factors, and neighbourhood context." *Criminology* 38 (2000): 1109–42. [CrossRef]

16. Victor Galea. *Impenn biex jitnaqqas l-illitterizmu f'Malta u Ghawdex*. Floriana: Directorate for Lifelong Learning, 2009.

17. Sandro Spiteri. "Prototype education action plan launched at Cospicua primary." *Times of Malta*, 29 March 2009. http://www.timesofmalta.com/articles/view/20090329/education/prototype-education-action-plan-launched-at-cospicua-primary.

18. JosAnn Cutajar. *Community Involvement and the Revitalization of Birgu*, 2008. Available online: http://www.euto.org/downloads/england/Community_Involvement_JosAnn_Cutajar.doc (accessed on 10 December 2008).

19. Jessica Camilleri. "The Impact of Prison on Inmates: A Sociological Investigation of Former Inmates at Corradino Correctional Facility, Malta." Masters in Sociology, Sociology Department, University of Malta, 2011.

20. Robert J. Sampson, Stephen W. Raudenbush, and Earls Felton. "Neighbourhoods and violent crime: a multilevel study of collective efficacy." *Science* 277 (1997): 918–24.

21. Saviour Formosa. Crimemalta. 20 January 2010. Available online: http://www.crimemalta.com (accessed on 20 January 2010).

22. Saviour Formosa. "The dark side of Maltese crime." *Malta Today*, 9 September 2009. http://archive.maltatoday.com.mt/2009/09/09/t7.html.

23. Beverly L. Stiles, Xiaoru Liu, and Howard B. Kaplan. "Relative deprivation and deviant adaptations: The mediating effects of negative self-feelings." *Journal of Research in Crime and Delinquency* 39 (2000): 64–90.

24. Steven Box. *Power, Crime and Mystification*. London: Routledge, 1983.

25. Hazel Croall. Social Class, Social Exclusion, Victims and Crime. In *Victims, Crime and Society*. Edited by Pamela Davies, Peter Francis, and Chris Greer. London: Sage Publications, 2007, pp. 50–77.

26. Alan Shiell, and Jenny Zhang. External Costs of Poverty: A Conservative Assessment. A Report to the United Way of Calgary and Area. 2004. Available online: http://tamarackcommunity.ca/downloads/vc/cal_costsofpoverty04.pdf (accessed on 10 July 2012).

27. Robert Agnew. "A General Strain Theory of Community Differences in Crime Rates." *Journal of Research in Crime and Delinquency* 36 (1999): 123–55.

28. Stephen W. Baron, Leslie W. Kennedy, and David R. FordeMale. "Street Youths' Conflict: The role of background, subcultural and situational factors." *Justice Quarterly* 18 (2001): 759–89.

29. Charis E. Kubrin, and Weitzer Ronald. "Retaliatory homicide: Concentrated disadvantage and neighbourhood culture." *Social Problems* 50 (2003): 157–80.

30. George A. Kaplan, Elsie R. Pamuk, John W. Lynch, Richard D. Cohen, and Jennifer L. Balfour. "Inequality in income and mortality in the United States: Analysis of mortality and potential pathways." *British Medical Journal* 312 (1996): 999–1003.

9

PARTICIPATORY SENSING-BASED SEMANTIC AND SPATIAL ANALYSIS OF URBAN EMERGENCY EVENTS USING MOBILE SOCIAL MEDIA

Zheng Xu[1,2*], Hui Zhang[1], Vijayan Sugumaran[3], Kim-Kwang Raymond Choo[4], Lin Mei[2] and Yiwei Zhu[5]

[1]Tsinghua University, Beijing, China

[2]The Third Research Institute of the Ministry of Public Security, Shanghai, China

[3]Oakland University, Rochester, Michigan 48309, USA.

[4]University of South Australia, Adelaide, Australia.

[5]Zhejiang Business Technology Institute, Ningbo, China.

ABSTRACT

With the advances of information communication technologies, it is critical to improve the efficiency and accuracy of emergency management systems through modern data processing techniques. Geographic information system (GIS) models and simulation capabilities are used to exercise response and recovery plans during non-disaster times. They help the decision-makers understand near real-time possibilities during an event. In this paper, a participatory sensing-based model for mining spatial information of urban emergency events is introduced. Firstly, basic definitions of the proposed method are given. Secondly, positive samples are selected to mine the spatial information of urban emergency events. Thirdly, location and GIS information are extracted from positive samples. At last, the real spatial information is determined based on address and GIS information. Moreover, this study explores data mining, statistical analysis, and semantic analysis methods to obtain valuable information on public opinion and requirements based on Chinese microblog data. Typhoon Chan-hom is used as an example. Semantic analysis on microblog data is conducted and high-frequency keywords in different provinces are extracted for different stages of the event. With the geo-tagged and time-tagged data, the collected microblog data can be classified into different categories. Correspondingly, public opinion and requirements can be obtained from the spatial and temporal perspectives to enhance situation awareness and help the government offer more effective assistance.

Keywords: Participatory Sensing, Urban emergency events, Social media, Urban computing

1 INTRODUCTION

With the advances of information communication technologies, such as cloud computing [1, 2], Internet of Things [3, 4], and big data [5, 6], it is critical to improve the efficiency and accuracy of emergency management systems through modern data processing techniques. The past decade has witnessed the tremendous technical advances in sensor networks, Internet/Web of Things, cloud computing, mobile/embedded computing, spatial/temporal data processing, and big data, and these technologies have provided new opportunities and solutions to emergency management. Data processing/analysis in emergency management is a typical big data scenario. Numerous sensors and monitoring devices continuously sample the states of the physical world, while the web data processing techniques make the Internet a big data repository which can reflect the states of the cyber world and the human world. The efficient processing of these data imposes a challenge to the data management community. It is important to develop advanced data management and data processing mechanisms to support disaster detection, disaster response and control, rescue resource planning and scheduling, and emergency commanding. The objective conditions, such as the lack of information, timely changeable situation, short time for decision-making, and serious consequences, bring a great challenge for the government emergency response. The extraction of accurate information is very important in emergency response. Social media provide a platform for data mining and information extraction.

Emergency management aims to develop strategies and establish operations to decrease the potential impact of unexpected events (i.e., human or natural disasters). By quick response and rescue, it saves human lives from the secondary disasters and enhances the stability of communities after disasters. Emergency management involves four stages: planning and mitigation, preparedness, response, and recovery. Geospatial applications (including geographic information system (GIS)) have been extensively used in each stage of emergency management. Decision-makers can utilize the geospatial information to develop planning and mitigation strategies. GIS models and simulation capabilities are used to exercise response and recovery plans during non-disaster times. They help the decision-makers understand near real-time possibilities during an event.

The recent evolution of the Internet has permitted an unprecedented increase in content created by non-specialist users, thanks to a reduction in technical barriers [7]. Web users can provide their geographic information through social media, such as Twitter[1] and Weibo[2]; these are now commonly referred to as volunteered geographic information (VGI), having a huge potential to engage citizens in place-based issues and provide significant, timely, and cost-effective source for geographer's and other spatially related fields of research and management [8]. Usually, the Weibo users can be "social sensors [9]." A social sensor is defined as an agent that provides information about its environment on a social network after interaction with other agents [10]. The sensing message from social sensors can be used during a live fire emergency or traffic updates. Studies on emergency response using social media data have drawn much academic attention. Existing studies mainly focus on three perspectives: (1) information spreading on social network; (2) geolocation information subtraction; and (3) semantic analysis. The information dissemination of social media can enhance timely situation awareness in a crisis situation, because they can report on-the-ground information from general public and allow the government to spread information to a wide group of people. The geolocation information of social media data plays an important role in emergency detection and quick response. The geolocation related to social media data included content-based location, posting location, and registration location, and they were used to estimate the geographical location of a crisis. The semantic analysis of twitter messages is also of significance in situation awareness and emergency response. It can help extract information on damages, request for assistance, casualties, or warning to enhance situation awareness and receive assistance during crisis.

In this paper, a participatory sensing-based model for mining spatial information of urban emergency events is introduced. Firstly, basic definitions of the proposed method are given. Secondly, positive samples are selected to mine the spatial information of urban emergency events. Thirdly, location and GIS information are extracted from positive samples. At last, the real spatial information is determined based on address and GIS information. Moreover, this study explores data mining, statistical analysis, and semantic analysis methods to obtain valuable information on public opinion and requirements based on Chinese microblog data. Typhoon Chan-hom is used as an example. Semantic analysis on microblog data is conducted, and high-frequency keywords in different provinces are extracted for different stages of the event. With the geo-tagged and time-tagged data, the collected microblog data can be classified into different categories. Correspondingly, public opinion and requirements can be obtained from the spatial and temporal perspectives to enhance situation awareness and help government offer more effective assistance.

The rest of the paper is organized as follows. In the next section, the related work is given. Section 3 presents the proposed model. Case studies on real data sets are conducted in Section 4. The application on the proposed method is given in Section 5. The last section gives the conclusion of our work.

2 RELATED WORKS

The early related work in this emerging research field has addressed the use of social media to generate map mashups to support collaborative real-time mapping [11] and give the overview of harvesting geospatial content [12]. Earle et al. [13] attempted assessing how fast tweeters reacted to the smaller (4.3 magnitude) and much more localized earthquake of Morgan Hill, CA, in March, 2009. Recently, the work of harvesting spatial information from the web has seen some activity in recent years. For example, it has been demonstrated that general purpose points of interest (POI) can be automatically derived from the users' map annotations [14] and vague geographic regions (e.g., Midlands or Middle West) delineated. Besides the texture content, georeferenced pictures from the photo-sharing website such as Flickr have been processed in terms of their density to show where the most famous landmarks are for a given location [15]. The Geospatial Exploratory Data Mining Web Agent that retrieves geographic information from web pages (related to outdoor activities) has also recently been discussed [16].

Recently, with the high-speed development of the social networks, such as Twitter and Weibo, many researchers have published their work of using the data from social networks including special events and localization of natural disasters [17]. Sakaki et al. [18] investigated the real-time nature of Twitter and put particular attention to event detection. The Twitter users are regarded as sensors. Their messages are used for detecting earthquake. A reporting system is developed for use in Japan by their proposed methods. Crooks et al. [19] thought Twitter as a distributed sensor system. They analyzed the spatial and temporal features of the Twitter feed activity responding to a 5.8-magnitude earthquake. Their experimental results argued that the Twitter users represent a hybrid form of a sensor system that allows for the identification and localization of the impact area of the event. Longueville et al. [20] used Twitter as a source of spatial-temporal information. By focusing on a real-life case of forest fire, they aimed to demonstrate its possible role to support emergency planning, risk, and damage assessment activities.

3 THE PROPOSED METHOD

In this section, the proposed method is illustrated. Firstly, basic definitions of the proposed method are given. Secondly, positive samples are selected to mine the spatial information of urban emergency events. Thirdly, the semantic analysis of the social media is given. Fourthly, location and GIS information are extracted from positive samples. At last, the real spatial information is determined based on address and GIS information.

3.1 Basic definitions

In this section, basic definitions of the proposed method are given. An event is something that happens at some specific time, and often some specific places [15]. In the proposed method, the spatial information of an urban emergency event can be mined since some messages sensing by social users have exact location information. An urban emergency event is defined as follows.

3.1.1 *Definition 1. Urban emergency event, e*

An urban emergency event e is a tubal {T, L}, where T and L is the life course and spatial of e.

3.1.2 *Definition 2. Words set of an urban emergency event e, W*

Words set consists of keywords of an urban emergency event. Usually, these words can be used to search messages of an urban emergency event. It is noted that elements in the words set are synonyms since social users may use different words to describe the same urban emergency event.

Crowdsourcing or participatory sensing may be a potential solution mining the spatial information of urban emergency events. The proposed method aims at collecting and analyzing the spatial information from social users. The social network can be seen as a sensor receiver. Usually, the social network users can be seen as social sensors. The proposed method is set as a hierarchical data model including three different layers. The different layers of the proposed method are illustrated in Fig. 1.

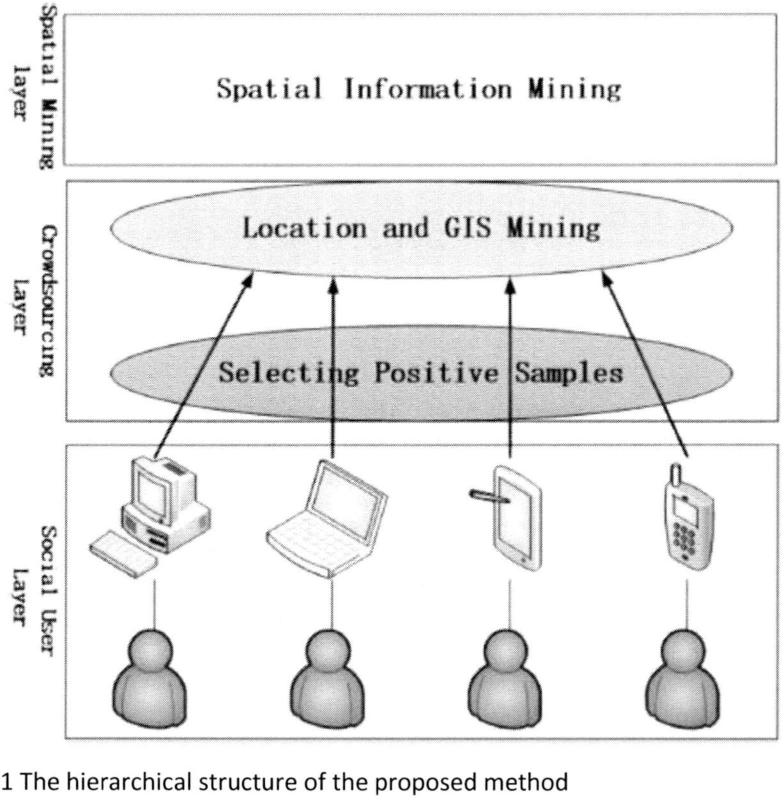

Fig. 1 The hierarchical structure of the proposed method

1. Social user layer. In this layer, the proposed method wants to collect the related data of urban emergency events. For example, if a user makes a message in Weibo about a fire occurrence, then the proposed method should collect this message.

2. Crowdsourcing layer. In this layer, address and GIS information is mined from messages from social users. Of course, location and GIS

information related to the same urban emergency events should be clustered.

3. Spatial information layer. In this layer, the spatial information of the urban emergency event is mined.

3.2 Selecting positive samples

3.2.1 *Definition 3. Positive samples of an urban emergency event e, PS*

Positive sample is the set of messages posted by social users, which can be used to mine the spatial information of an urban emergency event. For example, the message "A fire is happened at Jiangsu Road." can be seen as a positive sample. Based on common sense and our observations on real data, we have three heuristics to select positive samples from Weibo search results.

Heuristic 1: The Weibo message with the location or GIS information is prone to be a positive sample.

Since the location information is an important aspect for mining spatial information, the message with the location information can be thought as a positive sample. Different from the location information in Weibo message, the GIS information means the location of social users.

Heuristic 2: The Weibo message with the image or video information is prone to be a positive sample.

The image or video information is an important evidence to show the presence of social users. If a social user posts a message with the image or video of an urban emergency event, he is prone to be a witness of this event. Thus, he is also prone to post the accuracy spatial information of the urban emergency event.

Heuristic 3: The original Weibo message is prone to be a positive sample.

If a message is a forward message other than an original message, the social user is not a witness of the urban emergency event.

3.3 Semantic analysis

Semantic analysis aims at finding useful data from mass data to obtain valuable information or knowledge. Sina Weibo, one of the most popular microblogs in China, provides a platform for data mining. People express and record what they saw, heard, or experienced in the self-organized platform, and they also receive information from it. Sina Weibo is a huge information interaction center and the source of microblog data collection in this paper. The keyword-based search method was applied to collect microblog data through the Weibo API (an open interface of Sina Weibo). In the search, only microblog messages in Chinese language were collected and there were no geolocation constraints.

Text classification is to divide all the messages into several categories, and then semantic analysis is conducted in a certain category. The geo-tagged and time-tagged information of microblog messages was the standard for text classification. The collected microblog data were classified based on the geographic location of the posters. The number of microblog posts was counted in each province.

3.4 Mining location and GIS information

3.4.1 *Definition 4. The location information of an urban emergency event e, LI*

The location information can be extracted from the posted messages from positive samples. For example, in Fig. 2, a user post a message "I see a fire at Huaihai Road", the posted location information "Huaihai Road" can be extracted. In the proposed method, the location information detection is based on the Baidu Map, which can detect whether a word is a location name or not.

3.4.2 *Definition 5. The GIS information of an urban emergency event e, GI*

Unlike the location information, the GIS information is usually extracted from the check-in information from Weibo messages. The GIS information can be mined from the element of HTML page. For example, in Fig. 2, the GIS information of the message is "geo = 121.46, 31.22".

3.5 Determine spatial information

The spatial information of an urban emergency event usually consists of two aspects including the happening spatial information and the spread spatial information. The happening spatial information means the happening place of the urban emergency event. For example, in Fig. 2, the happening place of the fire is 358 Huaihai Road. The spread spatial information means the influential scope of the urban emergency event. For example, in Fig. 2, the influential place of the fire is Ruijing Road, which is shown in the check-in information. The influential place is usually equal to the check-in information of the social users. The check-in information means that the social users can see the happening emergency events in his place.

Fig. 2: The illustration example of location information. The message is "358 Huaihai Road is under fire". The check-in is "Ruijing 1 road, Luwan District, Shanghai"

4 SIMULATIONS ON REAL URBAN EMERGENCY EVENTS

4.1 The brief background of the Typhoon Chan-hom event

Typhoon Chan-hom hit Zhejiang province at 16:40 on July 11, 2015 as a category 2 storm. It was one of the strongest ones to hit the region this year and caused tremendous damages. More than one million people were evacuated in Zhejiang, and total economic losses in the province amounted to ¥8.86 billion (US$1.43 billion).

Typhoon Chan-hom formed in the northwest Pacific Ocean and was classified as a tropical depression on June 30, 2015. It was assigned the name Chan-hom when it upgraded to a tropical storm on July 1. With heavy storm and extreme rain, Chan-hom made landfall on Zhoushan, Zhejiang, on July 11, 2015 and the maximum sustained winds reached to 45 m/s. Thereafter, the storm accelerated north-northeast and moved over the Yellow Sea with degradation of its structure. Chan-hom was tracked each day based on the physical data mainly including GIS data, wind speed, rain rate, storm surge, and typhoon intensity.

4.2 The semantic analysis result

The keyword-based search method was applied to collect microblog data through the Weibo API (an open interface of Sina Weibo). The keywords included "Can Hong" (the Chinese of "Chan-hom"), "typhoon," and "Typhoon Chan-hom," and the period of data collection was from July 8, 2015 to July 15, 2015. In the search, only microblog messages in Chinese language were collected, and there were no geolocation constraints. After filtering the irrelevant data, the number of event-related microblog posts was 3321 with the geographic location of the posters and posting time attached. The number of microblog posts was counted in each province. The top four provinces were Zhejiang, Jiangsu, Shanghai, and Beijing, and their details were shown in Table 1. Based on the posting time, the collected data were classified into two categories: (1) before typhoon landfall (0:00 on July 8 to 16:40 on July 11) and (2) after typhoon landfall (16:40 on July 11 to 23:59 on July 15). The number of microblog posts at these two stages was 1154 and 2167, respectively, as shown in Table 2.

Table 1: Classification by geolocation (the top four provinces)

Provinces	**Zhejiang**	**Jiangsu**	**Shanghai**	**Beijing**
Number of microblog posts	1320	284	257	206

Table 2: Classification by posting time

Stages	**Before landfall**	**After landfall**
Period	July 8, 0:00– July 11, 16:40	July 11, 16:40–July 15, 23:59
Number of microblog posts	1154	2176

4.3 Risk perception and public opinion

Risk perception can be estimated by the number of social media messages to a certain degree. Risk perception was a driver to motivate a person to use social media during crisis. The scale that people use social media reflects people's risk perception during emergency. The number of microblog posts in different provinces was estimated based on the geolocation information. The risk distribution map was drawn based on the number in each province, as shown in Fig. 3. It was clear that the citizens' risk perception in some areas was relatively high, such as Beijing and some coastal provinces, i.e., Zhejiang, Jiangsu, and Shanghai. The closer away from the landfall place was, the higher the citizens' risk perception was. Beijing, as the capital of China, citizens there tends to pay more attention on crisis than people in other provinces.

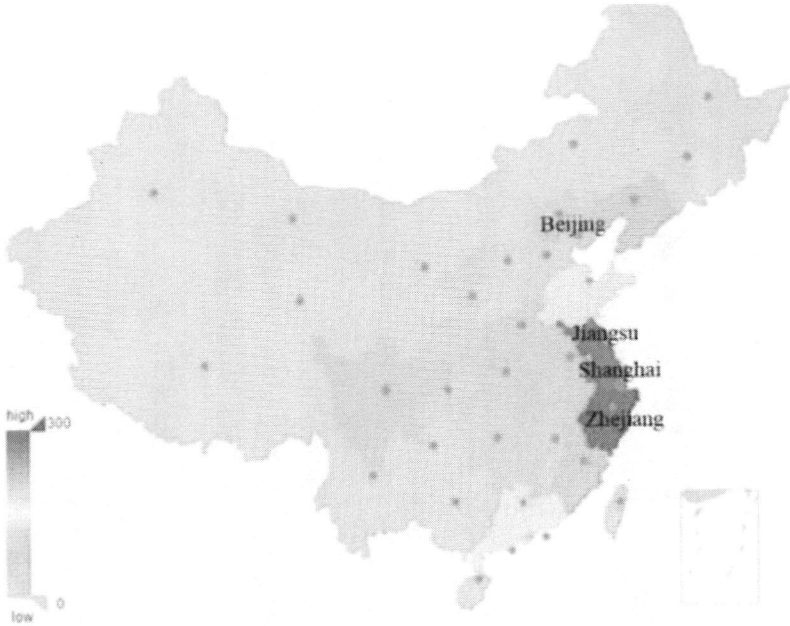

Fig. 3: Risk distribution map. In order to make the figure readable, the maximum of the range was 300 though the number of microblog posts in Zhejiang was 1320

The semantic analysis on the geo-tagged microblog data helped obtain public opinion from the spatial perspective. The assistance could be offered to where it was required. Keyword-frequency analysis and comparative analysis were conducted in three provinces where risk was relatively high. These provinces were Zhejiang, Shanghai, and Beijing. They were the representatives of the landfall place, the adjacent province, and the capital, respectively.

In this study, public opinion was extracted from the keywords whose frequencies were higher than 10. The number after a keyword represented its frequency. The comparative analysis of keywords distribution of these three provinces was shown in Fig. 4. The extracted keywords were classified into five categories based on their meanings. They were current situation, damage information, response, future situation, and citizens' requirements. The number of keywords describing current situation was the biggest of the five categories in all provinces. The keywords of Zhejiang mainly clustered in four categories: current situation, damage information, response, and citizens' requirements while Shanghai and Beijing were current situation, response, and future situation. It was clear that the number of keywords was bigger and their frequencies were higher in Zhejiang than in other provinces.

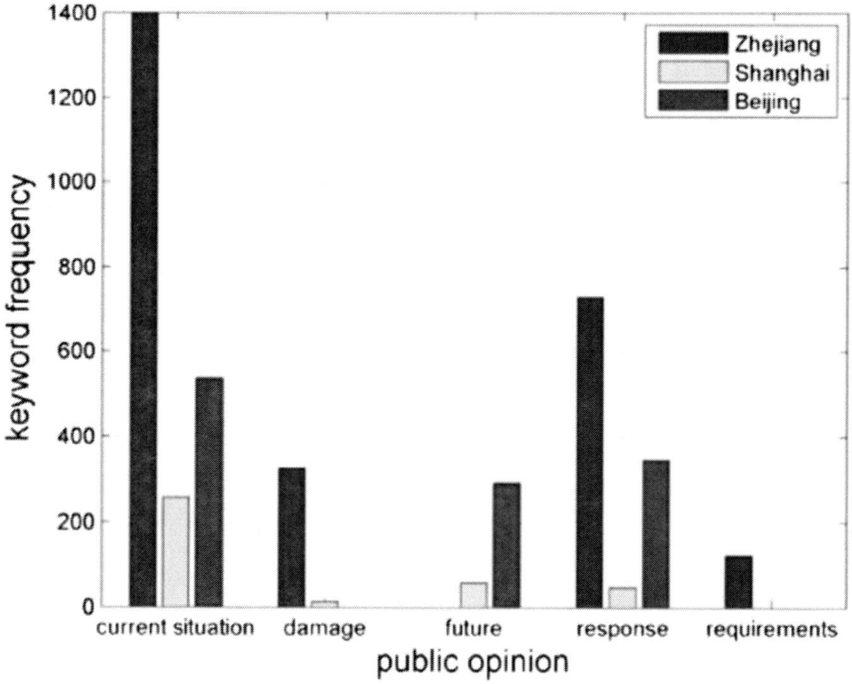

Fig. 4: Comparative analysis of keyword-frequency distribution in different provinces

The attention was paid more by the citizens in the place where a disaster happened than in other provinces. The citizens there concerned the damages and their needs while the citizens in other places wanted to know more on future situation, such as the possible track, intensity, tendency, and meteorology.

4.4 Spatial mining on the real explosion event

The *explosion* is also a major damage to the urban especially the big city. With the development of terrorism, the number and influence of the explosion event has become higher than before. For example, the "Boston Marathon" explosion event causes dozens of people dead. Thus, it is important to build the knowledge base of explosion emergency event. For example, the image or video of an explosion can be used by policeman to detect the suspect.

We select an explosion event that happened in 18:50 at "Taoyuan Apartment" on January 17, 2015 in Dalian. The proposed method is used to build the knowledge base of that event. The word "爆炸 (explosion)" is used to search in Weibo from 18:00 to 22:00. The search location is set as Dalian. Totally, 478 messages are returned. According to principles 1, 2, 3, and 4, 45 messages providing location information, check-in information, and image are selected as candidate messages. Figure 5 shows the timeline description of this case study. Only six timestamps are selected to build the timeline since they provide new keywords or locations. The red point in the map means the real location of that fire event. The red point with number means the GIS information of the different timestamps.

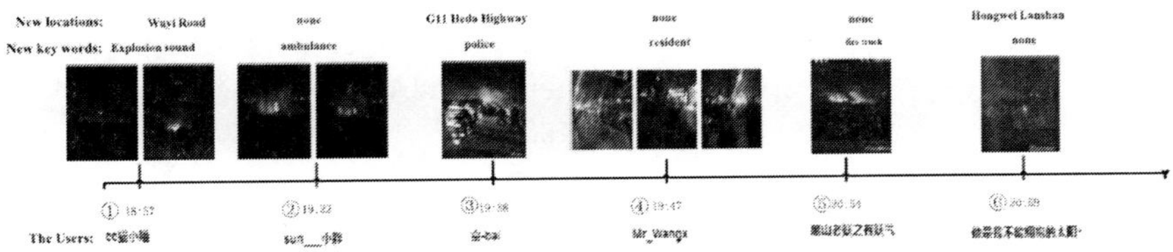

Fig. 5: The "explosion" case study using the proposed method

5 THE APPLICATIONS ON THE PROPOSED METHOD

In this section, a natural gas leakage monitoring system is developed, which composes sensor network and data collection system based on cloud storage platform. The sensor network consists both mobile and stationary sensors, which help to provide large-scale city areas monitoring with less sensors and provide more rapid and accurate leakage locating with GIS mapping.

The functions of concentration monitoring are as follows. (1) Concentration monitoring includes the provision of a methane concentration heat map. The original data collected by the mobile detectors and stationary detectors are merely a cluster of discrete locations that describe the relationship between the GPS information and methane concentration. Therefore, these data locations must be processed using a classic Gaussian plume diffusion model in order to obtain the concentration distribution of the entire monitored area. Then, these data are shown by means of distribution on a GIS map. (2) Concentration monitoring also includes mobile detectors. The real-time locations (latitude, longitude, and height) and methane concentrations are shown on the GIS map. In addition, every mobile detector is designated by a different color (red, orange, yellow, or green) according to the methane concentration. Red signifies the highest methane concentration and green the lowest. (3) Concentration monitoring includes stationary detectors. The real-time methane concentration and wind speed and direction obtained from each stationary detector are shown on the GIS map. The color designation method is similar to that for the mobile detectors. (4) The history data obtained from all the detectors can be queried via history methane concentration curves when the user clicks on a detector. The history data provide accuracy and reliability with regard to leakage warning. (5) Concentration monitoring includes the working state of the detectors. All the data for the citywide natural gas leakage monitoring system come from every detector, so the working state of the detectors determines whether the system is working normally. Therefore, it is necessary to monitor the working state of all the detectors. Each detector should periodically transfer a diagnostic report message to the control center. If the detector sends a faulty report or no report, then the detector is faulty and the maintenance staff must address such problems. (6) Concentration monitoring includes the level of warning. The real-time warning presents the security status of the monitored area. (7) Concentration monitoring includes a query detector. Users are able to query each detector according to the detector number and highlight it on the GIS map.

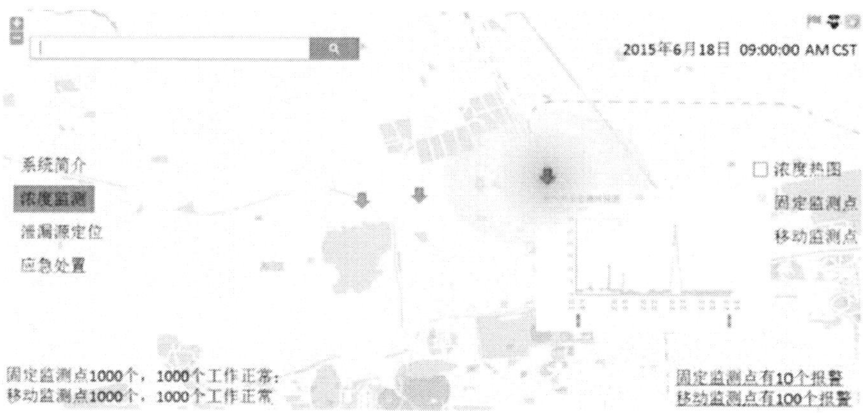

Fig. 6: Comparative analysis of keyword-frequency distribution in different provinces

By clicking on the leakage location, the relative position information will appear on the GIS map, including the district name and street names, as well as the locations of significant subjects, such as maintenance center, firefighter offices, hospitals, schools, and chemical plants. A message will be sent automatically to relevant offices, such as management office of gas companies, maintenance center, and emergency response office. Figure 6 shows examples of the system.

6 CONCLUSIONS

In this paper, a participatory sensing-based model for mining spatial information of urban emergency events is introduced. Firstly, basic definitions of the proposed method are given. Secondly, positive samples are selected to mine the spatial information of urban emergency events. Thirdly, location and GIS information are extracted from positive samples. At last, the real spatial information is determined based on address and GIS information. Moreover, this study explores data mining, statistical analysis, and semantic analysis methods to obtain valuable information on public opinion and requirements based on Chinese microblog data. Typhoon Chan-hom is used as an example. Semantic analysis on microblog data is conducted and high-frequency keywords in different provinces are extracted for different stages of the event. With the geo-tagged and time-tagged data, the collected microblog data can be classified into different categories. Correspondingly, public opinion and requirements can be obtained from the spatial and temporal perspectives to enhance situation awareness and help government offer more effective assistance.

NOTES

This paper is the extended version (50 % new content) of the conference paper accepted by EM-GIS2015.

ACKNOWLEDGEMENTS

This work was supported in part by the National Science and Technology Major Project under Grant 2013ZX01033002-003, in part by the National High Technology Research and Development Program of China (863 Program) under Grant 2013AA014603, in part by the National Science Foundation of China under Grant 61300202, in part by the China Postdoctoral Science Foundation under Grant 2014 M560085, the project of Shanghai Municipal Commission of

Economy and Information under Grant 12GA-19, and in part by the Science Foundation of Shanghai under Grant 13ZR1452900.

COMPETING INTERESTS

The authors declare that they have no competing interests.

AUTHORS' INFORMATION

Zheng Xu was born in Shanghai, China. He received Diploma and Ph.D. degrees from the School of Computing Engineering and Science, Shanghai University, Shanghai, in 2007 and 2012, respectively. He is currently working in the third research institute of ministry of public security and the postdoctoral in Tsinghua University, China. His current research interests include topic detection and tracking, semantic Web, and Web mining. He has authored or co-authored more than 70 publications including IEEE Trans. On Fuzzy Systems, IEEE Trans. On Automation Science and Engineering, IEEE Trans. On Cloud Computing, IEEE Trans. On Emerging Topics in Computing, and IEEE Trans. on Systems, Man, and Cybernetics-Part C.

Hui Zhang is a professor at the Department of Engineering Physics, Tsinghua University. Hui Zhang's primary research lies in the area of thermal/fluid science, crystal growth, disaster prevention and mitigation, and thermal spray coating. Prof. Zhang is now the PI of a major research program of NSFC focused on "study on the city's comprehensive risk assessment and several key issues in emergency response". He is one of the directors of the UTC-Tsinghua Joint Institute on Building Energy, Safety and Control System. He focuses on developing next-generation building safety technology through integrated information and sense systems, effective situation awareness, and evacuation strategy.

Vijayan Sugumaran is a professor of Management Information Systems and Chair of the Department of Decision and Information Sciences at Oakland University, Rochester, MI, USA. He received his Ph.D. in Information Technology at George Mason University, Fairfax, VA, USA. His research interests are in the areas of Big Data Analytics, Intelligent Information Technologies, Ontologies, and Semantic Web. He has published over 150 peer-reviewed articles and edited 12 books. He has published in top-tier journals, such as "Information Systems Research," "ACM Transactions on Database Systems," "IEEE Transactions on Engineering Management," "Communications of the ACM," "IEEE Transactions on Education," and "IEEE Software". Dr. Sugumaran is the editor-in-chief of the *International Journal of Intelligent Information Technologies*. He is the Chair of the Intelligent Agent and Multi-

Agent Systems mini-track for Americas Conference on Information Systems (AMCIS 1999–2016). He has served as the program co-chair for the International Conference on Applications of Natural Language to Information Systems (NLDB 2008, NLDB 2013 and NLDB 2016).

Kim-Kwang Raymond Choo received Ph.D. in Information Security in 2006 at Queensland University of Technology, Australia. He is currently a senior researcher at the University of South Australia. He was named one of the 10 Emerging Leaders in the Innovation category of The Weekend Australian Magazine/Microsoft's Next 100 series in 2009 and is the recipient of various awards including ESORICS 2015 Best Research Paper Award, Highly Commended Award in the 2014 "Best Chapter in a Book" Category by Australia New Zealand Policing Advisory Agency (ANZPAA), a Fulbright Scholarship in 2009, and a 2008 Australia Day Achievement Medallion.

Lin Mei received Ph.D. degree at Xi'an Jiaotong University, Xi'an, China, in 2000. He is a research fellow. From 2000 to 2006, he was a postdoctoral researcher at Fudan University, Shanghai, China; the University of Freiburg, Freiburg im Breisgau, Germany; and the German Research Center for Artificial Intelligence. He is currently the Director of the Technology R&D Center for the Internet of Things with the Third Research Institute of the Ministry of Public Security, China. He has published more than 40 papers. His research interests include computer vision, artificial intelligence, and big data processing.

Yiwei Zhu is an associate professor at Zhejiang Business Technology Institute, China. His current research interests include machine learning and data mining.

FOOTNOTES

1 www.twitter.com

2 www.weibo.com

REFERENCES

1. Z Xu et al., Knowle: a semantic link network based system for organizing large scale online news events. Futur. Gener. Comput. Syst. **43–44**, 40–50 (2015)

2. L Wang, J Tao et al., G-Hadoop: MapReduce across distributed data centers for data-intensive computing. Futur. Gener. Comput. Syst. **29**(3), 739–750 (2013)

3. Z Xu et al., Semantic based representing and organizing surveillance big data using video structural description technology. J. Syst Software **102**, 217–225 (2015)

4. Z Xu, X Luo, S Zhang, X Wei, L Mei, C Hu, Mining temporal explicit and implicit semantic relations between entities using web search engines. Futur. Gener. Comput. Syst. **37**, 468–477 (2014)

5. C. Hu, Z. Xu, et al, Semantic Link Network Based Model for Organizing Multimedia Big Data. IEEE Trans. Emerg. Top. Comput. **2**(3), 376–387 (2014)

6. X Luo, Z Xu, J Yu, X Chen, Building association link network for semantic link on web resources. IEEE Trans. Autom. Sci. Eng. **8**(3), 482–494 (2011)

7. Business Models for the Next Generation of Software. http://www.oreillynet.com/pub/a/oreilly/tim/news/2005/09/30/what-is-web-20.html (2005)

8. M.F. Goodchild, Citizens as Voluntary Sensors: Spatial Data Infrastructure in the World of Web 2.0. Int. J. Spatial Data Infrastructures Res. **20**, 24–32 (2007)

9. T. Sakaki, M. Okazaki, and Y. Matsuo. in *Proceedings of the 19th international World Wide Web conference*. Earthquake Shakes Twitter Users: Real-time Event Detection by Social Sensors, (2010), p. 851–860

10. V Krishnamurthy, H Vincent Poor, A tutorial on interactive sensing in social networks. IEEE Transac. Comput. Soc. System **1**(1), 3–21 (2014)

11. K Field, J O'Brien, Cartoblography: experiments in using and organizing the spatial context of micro-blogging. Trans. GIS **15**, 5–23 (2010)

12. A. Stefanidis, A. Crooks, J. Radzikowski, Harvesting ambient geospatial information from social media feeds. GeoJournal. **77**, 50–62 (2012)

13. P Earle, M Guy, R Buckmaster, C Ostrum, S Horvath, A Vaughan, OMG earthquake! Can Twitter improve earthquake response? Seismol. Res. Lett. **81**, 246–251 (2010)

14. L Mummidi, J Krumm, Discovering points of interest from users' map annotations. GeoJournal **72**(3–4), 215–227 (2008)

15. D. Crandall, L. Backstrom, D. Huttenlocher, J. Kleinberg, in *Proceedings of the 18th International World Wide Web Conference*. Mapping the World's Photos. (2009), p. 761–761

16. E. Pultar, M. Raubal, M.F. Goodchild, *in* Proceedings of the 16th ACM SIGSPATIAL International Conference on Advances in Geographic

Information Systems. GEDMWA: Geospatial Exploratory Data Mining Web Agent, (2008)

17. M Trusov, V Bodapati, E Bucklin, Determining influential users in internet social networks. J. Market. Res. **47**, 643–658 (2010)

18. T Sakaki, M Okazaki, Y Matsuo, Tweet analysis for real-time event detection and earthquake reporting system development. IEEE Trans. Knowl. Data Eng. **25**(4), 919–931 (2013)

19. A. Crooks, A. Croitoru, A. Stefanidis, and J. Radzikowski, Earthquake: Twitter as a Distributed Sensor System. (Transaction in GIS, 2012), pp. 1–26

20. B. Longueville, R. Smith, and G. Luraschi, in *Proceedings of the International Workshop on Location-Based Social Networks*. OMG, from here I can see the flames, a use case of mining location based social networks to acquire spatio-temporal data on forest fires, (2009), p. 73–80

10

ESTIMATION OF SOLAR ENERGY POTENTIAL FROM ROOFTOP OF PUNJAB GOVERNMENT SERVANTS COOPERATIVE HOUSING SOCIETY LAHORE USING GIS

Muhammad Luqman, Sajid Rashid Ahmad, Samiullah Khan, Usman Ahmad, Ahmad Raza, Farkhanda Akmal

Institute of Geology, University of the Punjab, Lahore, Pakistan

ABSTRACT

Pakistan has a severe electricity load shading problem. Government is trying to find out all ways for electricity generation. Alternative energy board is working to find out the energy potential using all alternative resources. Board has an objective to produce 9700 MW by 2030 to overcome load shading problem. A research was designed to find out the solar energy potential as an alternative source of energy from rooftops of residential areas in district Lahore. Punjab Governments servants housing society Lahore is selected. The society has minimum slope, aspect and shadows effects on the roofs. Also houses in PGSHS have same house structures and good town plan. A few portions of roofs are digitized to measure the available rooftop area for Photovoltaic panel's installation. GIS models are used to find out solar energy potential monthly as well as yearly for the year of 2014. The potential estimated is 39,613,072 kWh/year. The monthly total energy consumption of the society is 347,140 kWh which is only 11% of energy production from PV solar panels. As the estimated energy is 9 times than the energy demand of the society, extra energy can be used in local/national electricity transmission grid. Solar PV energy would be supplement to compensate energy shortfall in local area.

Keywords: Photovoltaic, Alternative Energy, Pakistan, CO_2 Emission, Rooftop

1. INTRODUCTION

On the earth, the biggest source of energy is the sun. Solar flux is more than many times of human need. Solar energy is a essential part of life. Without it, live would be impossible. Annual average solar energy on entire planet may produce 1.3 - 1.4 kW/m^2 electric energy [1] [2] . Intensity of solar flux depends upon geographic location, weather condition, altitude and diurnal variation. The minimum solar potential available worldwide is 0.06 kW/m^2 (~500 kWh/m^2 per year) to the highest flux of 0.25 kW/m^2 (~2200 kWh/m^2/year) in deserts areas of Africa and Australia [3] .

Aims and Objectives

The aim is to find out photovoltaic solar energy potential from a suitable site in Lahore City having maximum solar irradiance and minimum slop, shadow and aspect in rooftops using GIS technique.

This study proposes Geographic information system techniques and modeling to identify optimal place to estimate solar energy potential from the rooftop of the site in district Lahore. Optimal site in this study is consider that site which have very low topographic variation and very low slop, aspect and shadow affects at the rooftops of the buildings.

The objective is to estimate solar energy potential from rooftops of selected sites. The selected site would be residential area. The output generated energy would be consumed by these houses.

The second object of the study is to find out whether there is extra energy after the uses of rooftop owners of the site, to connect it to local electricity grid network. This energy would be utilized from other consumers.

This study also objects to find out how and how much carbon emission would be reduced by using solar energy technologies.

Energy short fall is a burning issue of Pakistan in present condition. Every field of life is facing energy shortfall problem especially residential area which is focus of our study. In June 2012, peak energy demand was 17,861 MW and peak energy generation capacity was 14,317 MW [4] . To compensate short fall of 3544 MW, consumers use other sources of energies, which are very costly, so solar energy is a solution for energy generation.

Arifeen [5] expressed in his article that electricity demand had increased by 80% in last 15 years. According to Pakistan Water and Power Development Authority (WAPDA), energy demand would be 40,000 MW by 2020 in Pakistan. An analysis of energy demand shows that there is 4.8% demand in last five years but now expected demand will be 8% - 10% per year, so 7% - 8% production is necessary to fulfill energy consumption in the future. For the energy production, some dams are being constructed in the country. Diamer Basha, Buniji and kohala Dams are some of them having capacity of 4500 MW, 7100 MW and 11 MW respectively. At present, about 300 small and mini hydropower plants are installed by the private and public sector in the northern hilly areas of Pakistan, supplying electricity to local consumers.

Unfortunately, all resources had not been utilized in Pakistan for energy production. There is lack of determination and management. In June 2011, a project for the construction of 12 small dams in different parts of the country was signed. Their completion date was 2013. However, not a single dam has been completed yet. The lack of planning is the big cause of energy crises. No practical work has been done for future energy demand. All plans died in paper work. Pakistan has a lot of resources for energy production. Hydro power energy is the cheapest energy, but no dams have been constructed from many years. Pakistan is granted by a huge amount of coal, but it is not utilized for energy production. A vast loss of electricity through transmission lines and theft produces an energy shortage. According to labor department, about 800,000 labors have lost their jobs. Many kids died in hospitals due to load shedding.

Electricity in Pakistan is being produced by using all conventional means of sources like hydro power, gas, oil and nuclear energy. Oil and gas are expensive sources. There was not any department for renewable energy till 2003; then, an autonomous board was made by government of Pakistan. This board is Alternative Energy Development Board (AEBD). The main job of this board is to find out all means of renewable energy and utilize them for energy production. This board has to achieve 5% share of power generation of total demand of electricity in country by 2030 [6] .

Solar radiations are 6000 times enough for total world energy demand and these radiations can fulfill dozen time energy demand of the world using current technologies [7] . Some land of Pakistan is located near tropical region of the earth. Long sun shine hours and high solar irradiance make Pakistan suitable for solar energy production.

The average global irradiance is about 200 - 250 Watt/m^2 per day over the horizontal surface. It is 1.9 - 2.3 MWh/m^2 per year [6] . The desert areas of Pakistan are Cholistan and Thar receiving a high amount of flux.

2. LITERATURE REVIEW

Photovoltaic panels' installation has many limitations all over the world. There are many models to estimate solar flux over a geographical area. All models have some parameters. All models have to face these limitations and challenges. Review of these challenges and limitations are as under:

Slope of rooftop is very important factor for installation of solar PV. Flat areas receive more radiations [8] - [11] . Rooftop having slop have directions. These directions on roof are called aspect of that roof. If these directions are not along the path of the sun then radiations will be affected [8] . Radiations fall with three ways on surface of the earth and roof. Those radiations which strictly direct on the rooftop or roof surface are called direct normal irradiance. The More the radiations more the energy produced. These radiations have a role for energy generation [12] -[14] . If modules aim to install in open land, then the project land should be near to roads for easy transportation of infrastructure. Project area close to roads results less cost of infrastructure construction, installation and maintenance [13] [14] . Distance of PV modules to transmission grid matters [10] [12] . Solar panel's efficiency may decrease due to fog, mist and dust [11] . Similarly environmental sensitive areas, Weather conditions, flood pathways, Dam sites and cultural and heritage sites also should consider for PV installation [11] [13] .

Johnson and Armanino [7] conduct a study in Marin County, California State to estimate solar PV potential from rooftop of the county. There is so much variance in topography of county. GIS models and other four type of data were used to estimate solar potential from rooftop. Shadows effect was the biggest challenge of the study. Rooftop areas were considered in this study rather than whole county land. As a result of the study there was 58 MW potential to produce energy from PV in commercial sector, 25 MW from parking lots and 13 MW form institutional facilities.

National Renewable energy laboratory (NREL) commissioned a study in U.S. at national level by working with Navigant Consulting Inc. (NCI) to estimate PV potential [15] . The result was more than 500 GW of energy potential from the rooftop by using PV.

National Renewable Energy Laboratory (NREL) completed a project in 2007 to find out solar and wind potential in Pakistan and Afghanistan. For estimating solar potential, European meteorological satellite data of three years' time span was used. Three types of irradiances, direct, diffuse and scattered were calculate at 10 KM and 40 KM grids for the entire country of Pakistan. Software named "GsT" is available on NREL website [16].

GIS Models for Solar Analysis

Many GIS models had been used for estimation of solar potential. Some models have good results and are being used presently. Different models use different kind of data and different operation systems and software models. Some of these models are as under:

A GIS based model, SolarFlux, was developed in ESRI software environment by Hetrick et al. [17] . Horizon shading, solar angle, surface orientation, atmospheric conditions and solar flux data is used in this software, to estimate solar energy potential of a specific area.

Another model by developed by Hofierka and Súri [18] for solar analysis in GIS Genasys software. This software was developed in Automation Markup language script. This algorithm works similar to Solar Flux model.

Mostly used software, "ArcGIS", is software of Environmental Sciences Research Institute (ESRI). It has a toolset for analyzing. In the toolset "Solar analyst" is one of those extensions. It was developed by the model of Fu and Rich [19] .

3. METHODOLOGY

3.1. Study Area

The Punjab Government Servants Housing Scheme Lahore was chosen due to its unique design and suitability for PV modules installation. Figure 1 is depicting surroundings of study area.

This society is centered at 31.40°N, 74.17°E and total area 524,692 m^2. As Lahore is a warm city so this society has a high irradiance have a high potential for PV energy generation. The average is temperature range between average high 30.8°C and average low 17.8°C (source: NOAA (1961-1990). In 2010, a total of 738 millimeters rain fall was recorded. The society is suitable for PV installation due to least topographic variability. Nearly all the roofs are of same height. There are four types of houses size. The classes are as:

Each Class has the same architecture design so all the roofs also have same design. The classes and number of houses are summarized in Table 1. This make easy and suitable place for PV installation. The study area was selected because Lahore has the very high PV potential across the world. The study was done using all the free data available.

Google Earth imagery is used for digitizing rooftops. High resolution Google Earth imagery was downloaded using Google Earth Pro of the date of March 18, 2014. Screen shot of the imagery is shown in Figure 2. Image has zero cloud cover and best for digitizing rooftops. This image was then georefred according to Google earth coordinates In Geographic coordinates system WGS84. ArcGIS 10.1 was used to Georef image.

Figure 1. Study area.

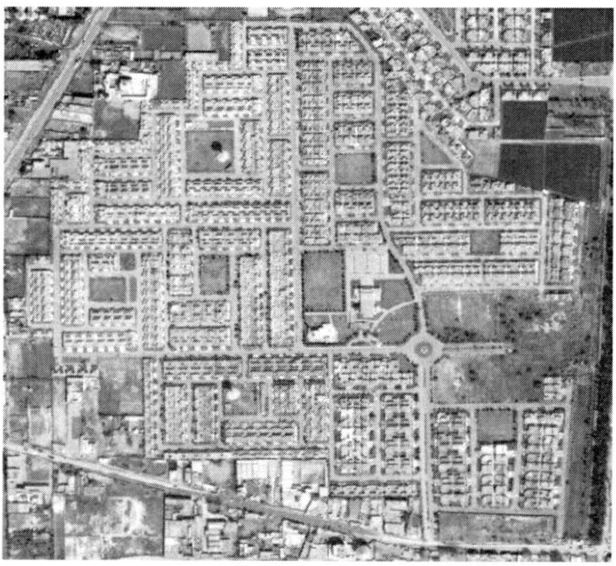

Figure 2. Aerial imagery of study area (Screen Shot of Google Earth Imagery).

Table 1. Study area buildings and types (source: http://www.pgshf.gop.pk/files/lahore.asp).

House Size	No. of Houses
5-Marla (126.47 M^2)	598
7-Marla (177.05 M^2)	334
10-Marla (252.93 M^2)	304
1-Kanal (505.86 M^2)	107

Google Earth imagery is used as a base map to digitize rooftops. The polygons were digitized at a zoom level of 1:250. Example of digitized rooftops is shown in Figure 3. Only a specific portion of rooftops is calculated. Not all the rooftop is digitized. Same size of area is selected from the houses of same size. Some Commercial buildings and school rooftop also digitized but a small portion not all available roof top. The mounts of rooftop are digitized so there is no shadow problem on selected area.

Only roofs were selected for PV installation during digitization which meets the following criteria:

As south, southeast and south west direction is best for PV installation, only buildings having required orientation were selected during digitization. There were no tilted roofs in study area. All buildings had flat surfaces, so orientation is not big issue for PV installation.

No shaded rooftop area was selected in this study. All the selected roofs receive direct solar radiations. Rooftop areas shaded by other buildings, structures or vegetation that affected the solar yield were not included in the study.

Those buildings having HVAC systems, air vents, chimneys etc., were not included in study area. No building had HVAC system nor chimneys etc. in study area.

DEM data was not helpful for slope and aspect calculations. A survey was conducted to collect elevation data manually for a few rooftops. Garmin GPS was used to collect elevation data. This Elevation data was entered in rooftops. Average elevation was entered to rooftops other than surveyed. Then these polygons were converted to raster to using polygon to raster tool in ArcGIS toolbox using elevation data as a field.

The survey was necessary to find out the average elevation of the study area for the calculation of annual incoming solar radiations. To understand geography, topography and rooftop shapes survey was conducted.

During survey it was noticed that nearly all the roofs have mounts on their roofs. These are only for security purpose that no one can enter in the house from down stairs. These mounts do not have any other use. So these mounts were selected for the PV installation. The mounts same as shown in Figure 4, were selected during survey of the study area.

Garmin GPS was used to find out elevation in the society. About 8 points were taken and average elevation was 202 Meters of study area. Forty bills were captured with a digital camera. As described above there are four types of houses in the study area having different sizes and different electricity load so it was necessary to collect data for all types.

Raster made from survey data was used as input to calculate slope and aspect in ArcGIS. Slope and aspect tools were used in ArcGIS 10.1 tool box. The results were aspect and slope as a raster. The aspect was calculated clockwise from north 0° - 359.9° in degrees measured and was shown in Figure 5. While the inclination was calculated from 0° - 90° in degrees as shown in Figure 6.

GIS model is a convenient way to generate solar radiation potential maps and there spatial relation to other data [19] . In this study, ESRI's ArcGIS's tool is used for GIS modeling in this study to calculate solar radiations. The main input for this tool is a Raster. This raster was made up of digitized roof polygons with average elevation as a field to convert it to raster using polygon to raster conversion tool from ArcGIS tool box.

Electricity rate and bill data was collected from Lahore Electric Supply Company (LESCO) website. LESCO provides electricity to entire division as well as Lahore city.

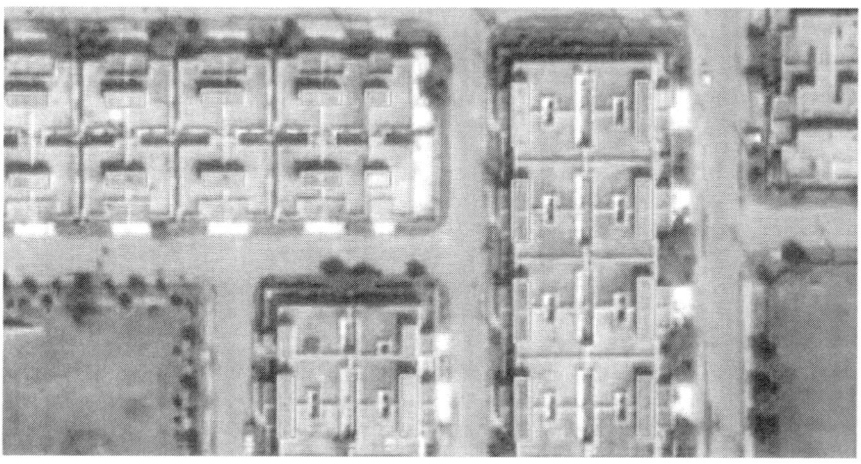

Figure 3. Sample set of rooftops (Screen Shot in ArcGIS).

Figure 4. Selection of roof mounts (Survey and Captured by Author).

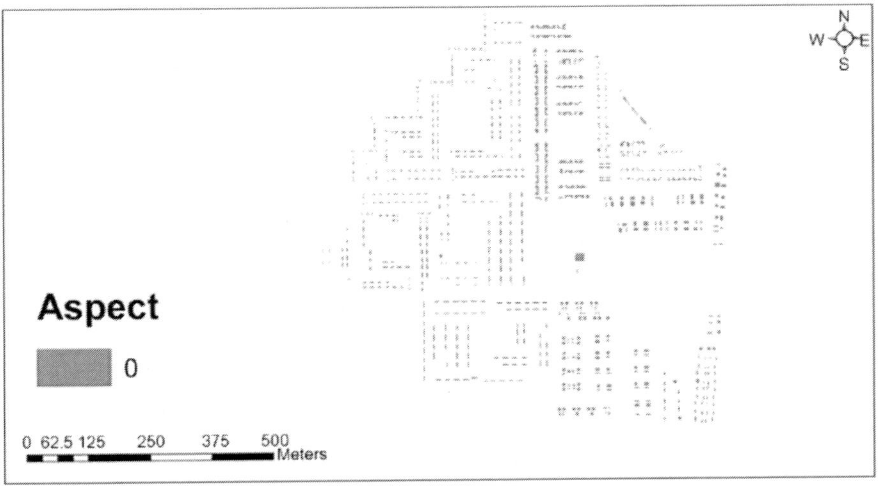

Figure 5. Map showing aspect of rooftops study area (Analysis in ArcGIS).

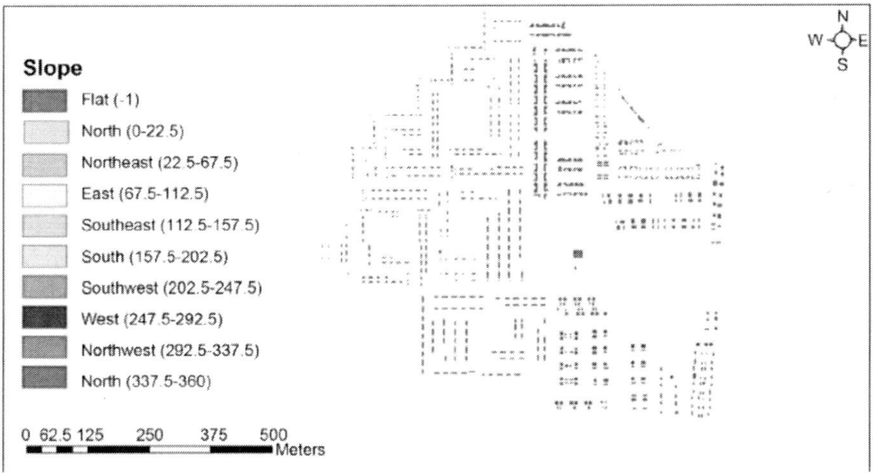

Figure 6. Map showing slop rooftops of study area (Analysis in ArcGIS).

3.2. CO_2 Emission Reduction

Carbon contributes about 80% of all greenhouse gases which causes global warming [20]. Carbon would causes half of climate change with in next century. Carbon emission is increasing day by day by the increasing demand of energy. Renewable energy technologies help to reduce Carbon emission. These technologies reduce less amount of carbon dioxide than fossil fuels. Solar energies produce low amount of energy as shown in Table 2.

Approximately 6.9×10^{-4} metric tons of CO_2 are emitted for the generation of 1 kWh of electricity.

1 kWh of solar energy = 6.9×10^{-4} metric tons of CO_2 reduction (1)

4. RESULTS

Average output solar energy potential in the study area for the year of 2014 is 1448 kWh/m^2/year as shown in Figure 7. All the roofs have same slop and aspect so every meter of the roofs of the study area would produce same amount of energy from solar flux.

After digitization rooftop area all the areas were calculated for individual building. A field was added to classify digitized polygons according to size of house. House size data was entered against digitized polygons.

To calculate terrain horizon, day length, global horizontal irradiance and temperature of the study area solargis (2014) was used. Solargis is a model to estimate solar radiations of a specific area all over the globe. Data of 4 geostationary satellite is utilized for estimation. Astronomical data and terrain data is also used by solargis in its model. In this study, solargis was used to find out some important factors which are necessary for our PV installation system.

Table 3 shows the all selectable area to install solar panel on the roofs. As generally all roof area is available but we suggest a small portion of roof for subjected purpose. House mounts are best suggested area.

To measure terrain horizon, solargis used STRMv2 data. In Figure 8, path of the sun is shown over the year of 2014 at study area. Three paths of sun are shown in the figure in three curves by using gray line. The lowest gray path of sun is during winter solstices, middle curve during equinox and top curve during summer solstices.

The sun curves shows that sun have low horizon during winters and high horizon in summers at the study area. Also all the maximum energy falls in south aspect of the horizon.

The zenith angle is the angle between the sun and the vertical perpendicular drawn upon a particular area. The zenith angle is like elevation angle but it is measured from the vertical rather than from the horizontal, thus making the zenith angle = 90°-elevation of sun from horizon.

It is clear than the sun has very low zenith angle during summer solstice and high zenith angle at vernal equinox. Solar zenith angle is also related do day time. The greater the zenith angles the lesser the day time hour and vice versa as it is clear in Figure 9.

Global solar irradiations are total radiations on the horizontal surface on earth. These radiations are sum of all radiations like direct normal irradiance, diffuse irradiance and ground reflected radiations. Ground reflected radiations are negligible. Both diffuse and direct radiations are shown in Figure 10 of study area. These are maximum in May, and minimum the month of January. The more the global irradiations the more the electric power generation. It is clear there would be maximum power generation in May and minimum in January.

Air temperature at 2 meters (TEMP) data was gained and calculated by solargis from NOAA NCEP data sources. Air temperature is available from solargis for the period from January 1994 to the present time.

Site: Lahore, Pakistan, latitude/longitude: 31.4052°N/74.1661°W

PV system: 1.0 kWp, crystalline silicon, fixed free

Azimuthal angle: 180° (south)

Inclination: 30

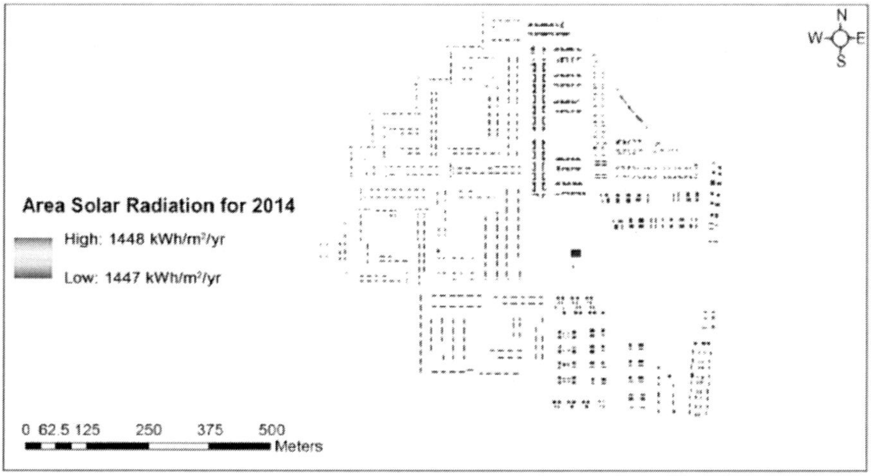

Figure 7. Annual solar radiation calculated for 2014 (Analysis in ArcGIS)

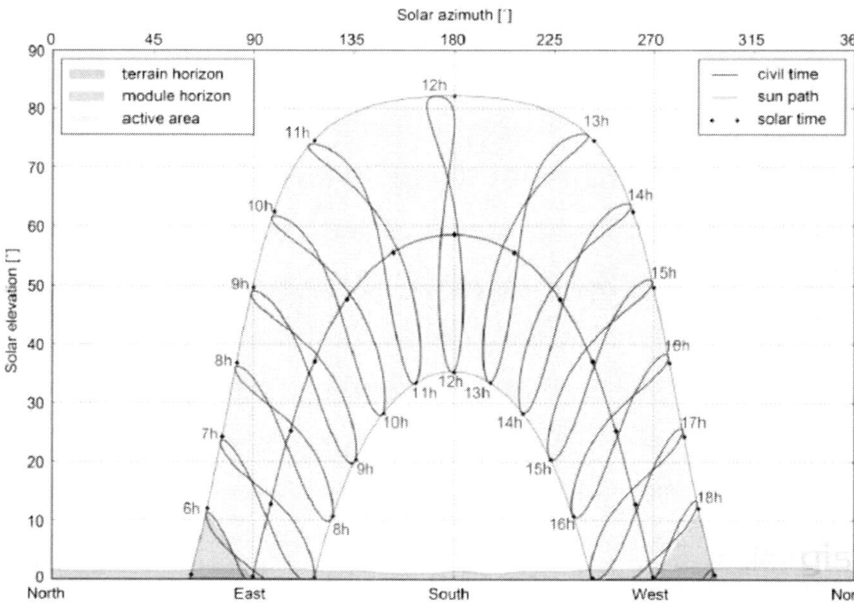

Figure 8. Terrain horizon at punjab government servants housing society Lahore (Source: solargis).

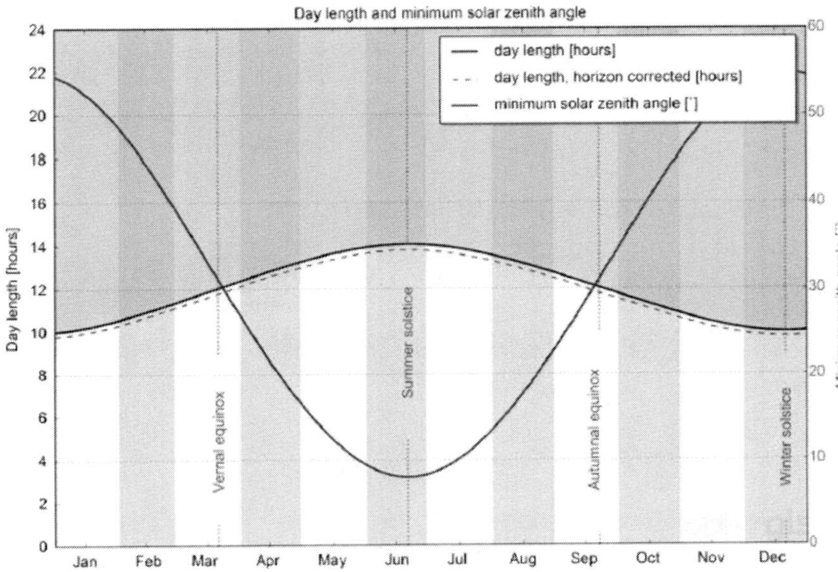

Figure 9. Day length and zenith angle at PGSHS Lahore (Source: solargis).

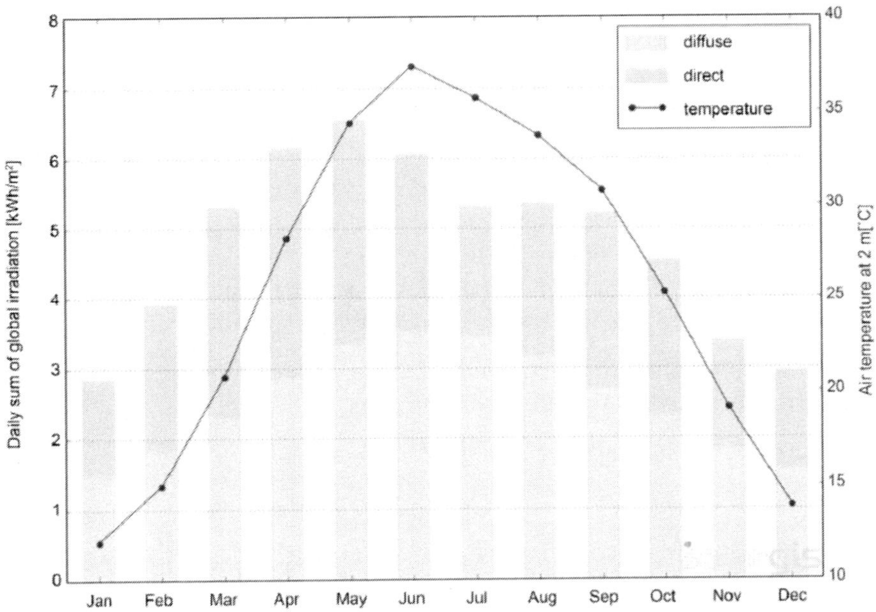

Figure 10. Global horizontal irradiation and air temperature at Punjab government servants housing society Lahore (Source: solargis).

Table 2. Carbon emissions from different sources of energy [21] .

Energy Source	Fossil Fuels	Solar	Wind	Hydro	Biogas
Carbon emissions (g/kWh generated)	504	99	10.2	10	11

Table 3. Suitable rooftop area by general building type.

Sr. #	House Size	Average Digitized Roof Area m² Per House	Total Digitized Roof Area m² All Houses	Percentage of Total Area
1	5-Marla (126.47 M²)	19	6674	1.27%
2	7-Marla (177.05 M²)	29	5089	0.97%
3	10-Marla (252.93 M²)	31	4860	0.93%
4	1-Kanal (505.86 M²)	37	10,212	1.94%
5	Other	60	541	0.10%
	Total Area for PV installation		27,376 m²	

If all suitable and estimated rooftop area is installed with solar PV panels the total power generation from Class D, 5 Marla houses would be 804,773 kWh/m^2 monthly and 9,657,278 kWh/m^2 annually. Similarly 7,367,783 kWh/m^2 would be produced from 7 Marla, 7,032,420 kWh/m^2 from 10 Marla, 14,776,764 kWh/m^2 from 1 Kanal and 782,827 kWh/m^2 would be produced annually. Table 4 shows the detail for the potential of each class and total area.

The average electricity is calculated from 1 year data collected for 12 houses in each class from LESCO website as shown in Table 5. 192 kWh energy is consumed in 5 Marla house, 261 kWh in 7 Marla, 305 kWh in 10 Marla and 490 kWh in 1 Kanal House per month.

Table 6 estimates the total energy consumption from all houses. These results are average. 5 Marla houses Class have highest energy consumption which is 114,816 kWh per month. 7 Marla houses have demand of 87,174 kWh, 10 Marla 92,720 and 1 Kanal demands 52,430 kWh per month.

Approximately 6.9 × 10^{-4} metric tons of CO_2 are emitted for the generation of 1 kWh of electricity (EPA 2005).

So by using Equation 3.1:

1 kWh of solar energy = 6.9 × 10^{-4} metric tons of CO_2 reduction

39,613,072 kWh of solar energy = 39,613,072 × 6.9 × 10^{-4} metric tons of CO_2 reduction

= 273,330,197 × 10^{-4} metric tons of CO_2 reduction

= 27,333 Metric tons of CO_2 reduction

5. DISCUSSION

The reason for more energy production from 5 Marla houses comparing to 7 and 10 Marla is, there are more number of houses in 5 Marla class. 1 Kanal houses have maximum energy generation potential because these houses occupy more land. Fifth class E consist of commercial, school and community center buildings, these have high potential because these building have more rooftop space available. Community center and school's all rooftop may be used for PV installation because these roof have no other use and fee. The study area has more energy generation potential than demand. In the long summer days when the energy short fall reaches its peak, solar energy generation would also be at its peak. Furthermore the house owners would get minimum electric bills instead they would get extra income by investing their extra energy in national grid. If all the roof area would utilized

Table 4. Solar power potential estimates for PGSHS Lahore.

Class	House Size	Total Estimated Roof Area All Houses	Monthly Potential Total Area kWh	Yearly Potential Total Area kWh
Class D	5-Marla (126.47 M^2)	6674	804,773	9,657,278
Class C	7-Marla (177.05 M^2)	5089	613,649	7,363,783
Class B	10-Marla (252.93 M^2)	4860	586,035	7,032,420
Class A	1-Kanal (505.86 M^2)	10,212	1,231,397	14,776,764
Class E	Other	541	65,236	782,827
	Total	27,376	3,301,090	39,613,072

Table 5. Electricity consumption of individual house at PGSHS.

Sr. #	House Size	Average Estimated Electricity Consumption per house per Month kWh	Monthly Average Potential per house type kWh
1	5-Marla (126.47 M^2)	192	1346
2	7-Marla (177.05 M^2)	261	1837
3	10-Marla (252.93 M^2)	305	1928
4	1-Kanal (505.86 M^2)	490	11,508

Table 6. Electricity consumption of all houses at PGSHS (Source: LESCO).

Class	House Size	Total Estimated Electricity Consumption All houses per Month kWh	Monthly Total Potential all houses house type kWh	Percentage consumption
Class D	5-Marla (126.47 M^2)	114,816	804,773	14%
Class C	7-Marla (177.05 M^2)	87,174	613,649	14%
Class B	10-Marla (252.93 M^2)	92,720	586,035	16%
Class A	1-Kanal (505.86 M^2)	52,430	1,231,397	4%
	Total	347,140	3,235,854	11%

the energy would be huge from the study area. A financial model would be proposed for the investment in the study area by different stake holders or energy companies.

6. CONCLUSIONS

Though small parts of roofs are selected, the energy production is very high. Each 5 Marla house would produce average 1346 kWh energy per month and 7 Marla would produce 1837 kWh per month. 1928 kWh and 11,508 kWh of energy would be produced by 10 Marla and 1 Kanal houses per month respectively.

This study sought out PV potential from PGSHS Lahore's building's rooftops. This Study provides an overview of PV energy potential. There were two main outputs from this study: 1) the high resolution incoming solar radiations data in study area, and 2) total rooftop area available for PV installation. After studying and analysis, we see that the total photovoltaic potential for electricity generation from the rooftops of PGSHS Lahore is approximately 399,613,072 kWh annually. Comparing with energy consumption of society, this would be approximately 11% of total energy consumption of society. There will be more energy generation if we utilize all rooftop areas.

There is 4.01 $kWh/m^2/day$ energy potential in the study area. According to NREL scale for energy potential, study area lies in good resource potential.

It is not possible to calculate exact capacity for proposed energy generation. We only can estimate from the calculated rooftop area. Many steps were taken to estimate rooftop area by digitations. High resolution imagery was used to digitize rooftops. A field survey was conducted to ensure whether all roofs were flat or tilted. Shadow effect of one roof on another is also avoided by only digitizing mount areas. And minimum area of roof is utilized.

Muhammad Luqman, Sajid Rashid Ahmad, Samiullah Khan, Usman Ahmad, Ahmad Raza, Farkhanda Akmal Total 1443 parcels were digitized from residential, commercial and school buildings in the study area. For exact calculations, all the rooftops were digitized rather than some samples. All the roofs are the same in the same class of houses. Nearly the same area is selected from all rooftops of the same class. From GPS survey, average elevation was calculated in the study area. This helps in estimating the solar radiations potential for all society. There was no slop and aspect in study area's rooftops, so radiations were the same on all roofs and calculation was not complex. A great care was taken during survey, during digitizing rooftops and during analysis.

REFERENCES

1. Rogner, H.-H., Barthel, F., Cabrera, M., Faaij, A., Giroux, M., Hall, D., et al. (2000) Energy Resources. In: Goldemberg, J., Ed., World Energy Assessment: Energy and the Challenge of Sustainability, UNDP, New York, 136-171.

2. Sustainable Energy Utility Task Force (2008) The Sustainable Energy Utility: A Delaware First. A Report to the Delaware State Legislature by the Sustainable Energy Utility Task Force. http://www.seu-de.org/docs/SEU_Final_Report.pdf

3. De Vries, B.J.M., van Vuuren, D.P. and Hoogwijk, M.M. (2007) Renewable Energy Sources: Their Global Potential for the First-Half of the 21st Century at a Global Level: An Integrated Approach. Energy Policy, 35, 2590-2610. http://dx.doi.org/10.1016/j.enpol.2006.09.002

4. PEPCO (2012) View Power Situations. http://www.pepco.gov.pk /pow_situation.php

5. Arifeen, M. (2013) Can Pakistan Meet Its Energy Needs? The Nation Newspaper, 7. http://nation.com.pk/columns/19-Jul-2013/can-pakistan- meet-its-energy-needs

6. Khalil, M.S. Renewable Energy in Pakistan: Status and Trends. Pakistan Alternative Energy Development Board, Prime Minister's Secretariat.

7. Johnson, G. and Armanino, D. (2004) GIS Tools for Community Development Applications. Paper Presented at American Solar Energy Society (ASES) Solar Power 2004 Conference, San Francisco, CA.

8. Bravo, J.D., Casals, X.G. and Pascua, I.P. (2007) GIS Approach to the Definition of Capacity and Generation Ceilings of Renewable Energy Technologies. Energy Policy, 35, 4879-4892. http://dx.doi.org/10.1016/j.enpol.2007.04.025

9. Pletka, R., Block, S., Cummer, K., Gilton, K.R.O., Roush, B., Stoddard, L., Tilley, S., Woodward, D. and Hunsaker, M. (2007) Arizona Renewable Energy Assessment. Black & Veatch Corporation, Overland Park. http://www.energy.ca.gov/reti/documents/2007-09_AZ_Renewable_Energy_Assessment.pdf

10. Broesamle, H., Mannstein, H., Schillings, C. and Trieb, F. (2001) Assessment of Solar Electricity Potentials in North Africa Based on Satellite Data and a Geographic Information System. Solar Energy, 70, 1-12. http://dx.doi.org/10.1016/S0038-092X(00)00126-2

11. Charabi, Y. and Gastli, A. (2011) PV Site Suitability in Oman Analysis Using GIS-Based Spatial Fuzzy Multi-Criteria Evaluation. Renewable

Energy, 36, 2554-2561. http://dx.doi.org/10.1016 /j.renene. 2010.10.037

12. Fluri, T.P. (2009) The Potential of Concentrating Solar Power in South Africa. Energy Policy, 37, 5075-5080. http://dx.doi.org/ 10.1016/ j.enpol.2009.07.017

13. Clifton, J. and Boruff, B.J. (2010) Assessing the Potential for Concentrated Solar Power Development in Rural Austra- lia. Energy Policy, 38, 5272-5280. http://dx.doi.org/10.1016/j.enpol.2010.05.036

14. Dawson, L. and Schlyter, P. (2012) Less Is More: Strategic Scale Site Suitability for Concentrated Solar Thermal Po- wer in Western Australia. Energy Policy, 47, 91-101. http://dx.doi.org /10.1016/j.enpol.2012.04.025

15. Paidipati, J., Frantzis, L., Sawyer, H. and Kurrasch, A. (2008) Rooftop PV Market Penetration Scenarios. Subcontract Report for NREL.

16. Perezi, R., Schlemmer, J., Moore, K. and George, R. (2007) Satellite-Derived Resource Assessment in Afghanistan & Pakistan in Support of the USAID South Asia Regional Initiative. Subcontract Report for NREL.

17. Hetrick, W.A., Rich, P.M., Barnes, F.J. and Weiss, S.B. (1993) GIS-Based Solar Radiation Flux Models. American So- ciety for Photogrammetry and Remote Sensing Technical Papers 3, GIS. Photogrammetry and Modeling, 3, 132-143. http://professorpaul.com/publications /hetrick_et_al_1993_asprs.pdf

18. Hofierka, J. and Súri, M. (2002) The Solar Radiation Model for Open Source GIS: Implementation and Applications. Proceedings of the Open Source GIS-GRASS Users Conference 2002, Trento, 11-13 September 2002. http://www.ing.unitn.it/~grass/conferences/ GRASS2002/proceedings/proceedings/pdfs/Hofierka_Jaroslav.pdf

19. Fu, P. and Rich, P. (1999) Design and Implementation of the Solar Analyst: An Arc View Extension for Modeling Solar Radiation at Landscape Scales. Proceedings of the Nineteenth Annual ESRI User Conference, San Diego. http://professorpaul.com/publications /fu_rich_1999_esri.pdf

20. Lashof, D.A. and Ahuja, D.R. (1990) Relative Contributions of Greenhouse Gas Emissions to Global Warming. Nature, 344, 529-531. http://dx.doi.org/10.1038/344529a0

21. Martin, P. (2006) Dynamic Life Cycle Assessment (LCA) of Renewable Energy Technologies. Renewable Energy, 31, 55-71. http://dx.doi.org/10.1016/j.renene.2005.03.002

11

A GIS-BASED FRAMEWORK FOR REAL-TIME DEBRIS-FLOW HAZARD ASSESSMENT FOR EXPRESSWAYS IN KOREA

Han-Saem Kim[1], Choong-Ki Chung[2], Sang-Rae Kim[3], Kyung-Suk Kim[4]

[1] Earthquake Research Center, Korea Institute of Geoscience and Mineral Resources, 124 Gwahak-ro, Daejeon, Korea

[2] Department of Civil and Environmental Engineering, Seoul National University, 1 Gwanak-ro, Seoul, Korea

[3] Department of Civil and Environmental Engineering, Northwestern University, 633 Clark St, Evanston, USA

[4] Expressway and Transportation Research Institute, Korea Expressway Corporation, 922 Dongbu-daero, Hwaseong, Korea

ABSTRACT

Debris flows caused by heavy rainfall in mountain areas near expressways lead to severe social and economic losses and sometimes result in casualties. Therefore, the development of a real-time system for debris-flow hazard assessment is necessary to provide preliminary information for rapid decision making about evacuations or restoration measures, as well as to prevent secondary disasters caused by debris flows. Recently, various map-based approaches have been proposed using multi-attribute criteria and assessment methods for debris-flow susceptibilities. For the macrozonation of debris-flow hazard at a national scale, a simplified method such as the Korea Expressway Corporation (KEC) debris-flow hazard assessment method can be applied for systematic analysis based on geographic information systems (GIS) and monitoring networks. In this study, a GIS-based framework of real-time debris-flow hazard assessment for expressway sections is proposed based on the KEC debris-flow hazard assessment method. First, the KEC-based method was standardized in a systematic fashion using ArcGIS, enabling the objective and quantitative acquisition of various attribute datasets. The quantification of rainfall criteria also was considered. A safety management system for debris-flow hazard was developed based on the GIS platform. Finally, the method was applied and verified on three expressway sections in Korea. The grading standard for each individual influencing attribute was subsequently modified to more accurately assess the debris-flow hazards.

Keywords: Debris-flow hazard, Expressway management, GIS, Korea, Real-time hazard assessment

1 INTRODUCTION

Landslides, and in particular debris flows, are one of the most damaging natural hazards in mountainous terrain that experiences heavy torrential rainfall. Rainfall-triggered landslides are a recurring problem in Korea due to a combination of mountainous terrain, a shallow layer of alluvial soil, and associated weather conditions (Hungr et al. 2001; Park et al. 2013). "Flowage" or "flow" is the term used to describe the downslope movement of unconsolidated material in which particles move about and mix within the mass, such as earthflows, debris flows, or avalanches. To avoid confusion resulting from the interchangeable use of terminology, the term "debris flow" in this study refers to the fluid mixture of rocks, sand, mud, and water that is intermediate between a landslide and a water flood (Feusto et al. 1999; Keller and DeVecchio 2008).

Debris-flow damage includes loss of human life, destruction of facilities, and damage to roads, pipelines, and vehicles (Jakob and Hungr 2005). Until now, only post-event repairs and protective infrastructure improvements have generally been executed after debris-flow occurrences. Recently, an increase has occurred in Korea in both the number of occurrences of and the costs for countermeasures against debris flows. In order to sufficiently manage and protect expressway sections and facilities that are exposed to debris-flow occurrences, a method to assess the hazard of debris flows during certain rainfall events at a regional scale is needed. Infiltration of prolonged, intense rainfall, which causes soil saturation and a temporary increase in pore water pressure, is the mechanism by which most shallow landslides, and more specifically debris flows, are generated during rainstorms (Iverson 2000; Keller and DeVecchio 2008). Because Korea is a region with heavy, concentrated, torrential rainfall in the summer, most of the debris flows occur during the rainy summer periods in mountainous regions. Due to climate change and other environmental factors, the intensity and frequency of rainfall events are increasing. This increase results in a greater number of debris-flow occurrences and higher restoration costs than in the past. Consequently, more debris-flow countermeasures are now necessary for the safety and preservation of human lives and infrastructure.

Assessment of landslides, including debris flows, has been carried out with geographic information systems (GIS) techniques combined with statistical analyses and physical-based approaches by various researchers such as Carrara et al. (1999), Dai and Lee (2002), Ohlmacher and Davis (2003), Ayalew et al. (2004), Wang et al. (2008), and Kritikos and Davies (2015). In spite of the importance of GIS approaches, there are few techniques for debris-flow zonation and geodatabase analysis, and there is no practical and standardized debris-flow mapping method using GIS (Paulín et al. 2014). Through comparison and review of existing studies on the factors that influence debris flow and appropriate assessment methods, the Korea Expressway Corporation (KEC) debris-flow hazard assessment method was developed as a fundamental assessment tool (Korea Expressway Corporation 2009). This eclectic method focuses on the probable occurrence of road structure hazards in a highway system. The method can be quantitatively and objectively performed in a simple manner by using documents such as numerical maps and expressway design files. By drawing on existing databases, the method minimizes the need for tiresome field investigations in countless potential debris-flow occurrence regions in vast areas. Above all, in order to immediately assess rainfall-induced debris flows, rainfall criteria were utilized to determine fundamental trigger values and indices for the severity level of debris flows.

According to the KEC method, zonation of debris-flow hazard is evaluated through two indices: susceptibility value and vulnerability value. The susceptibility value indicates the likelihood of debris-flow occurrence, and can be estimated with the topography information available on target locations. The vulnerability value is used to represent whether an occurred debris flow will damage or have an impact on specific expressway sections, and can be assessed from the capacity of the drainage facility and the roadside margin area for sedimentation of debris-flow materials before they reach expressway structures. Determination of these influencing factors is achieved through the use of digital maps and expressway design files. The calculated susceptibility value and vulnerability value are used to indicate a single hazard class (Fig. 1), which represents the likelihood of damage by debris-flow events under a given rainfall intensity.

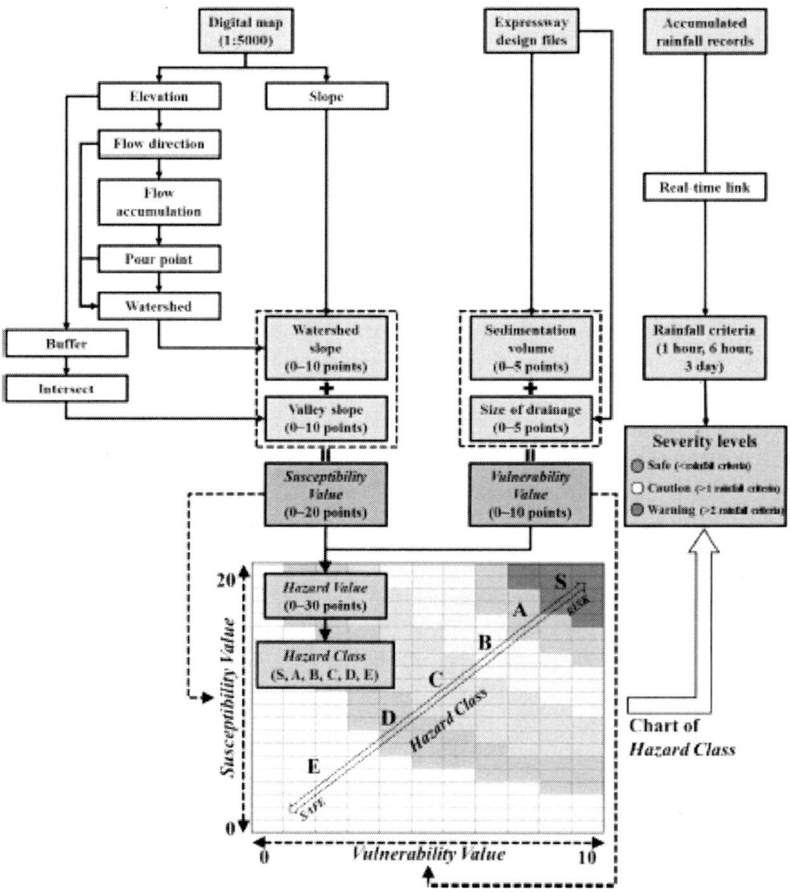

Fig. 1: Real-time debris-flow hazard assessment framework for expressways

In this study, the GIS-based framework of real-time debris-flow hazard assessment for expressway sections was proposed based on the KEC debris-flow hazard assessment method. To standardize the KEC method, a systematic sequence for the acquisition of attribute values was proposed using Arcgis 10.2 (Esri 2006). Then to consider the real-time link with the Automatic Weather Station (AWS) network in the Korean Meteorological Administration,[1] the rainfall recurrence periods were quantified as accumulated rainfall using the trigger values of past debris-flow occurrence events. Following the optimization of the KEC debris-flow hazard assessment method, a safety management system for debris-flow hazards was developed based on the ArcGIS platform. This system consists of a database (DB) and four systematic submodules: the input module, the debris-flow hazard assessment module, decision-making module of debris-flow countermeasure, and the real-time prediction module of debris-flow hazard, which is linked to the rainfall monitoring network. Finally, an applicability evaluation of the framework was carried out on three expressway sections of the Korean highway system that have suffered damage due to debris-flow occurrences during heavy rainfall events in 2005 and 2006: the Pyeongchang area of the Yeongdong Expressway, the Deogyu Mountain area of the Daejun-Jinju Expressway, and the Juksan-Geochang area of the 88 Expressway. The reliability of the assessment method was investigated by comparing the actual debris-flow occurrence and nonoccurrence cases.

2 LITERATURE REVIEW OF DEBRIS-FLOW HAZARD ASSESSMENT

A review of methods for debris-flow hazard assessment, including those by Dai and Lee (2002), Lin et al. (2002), Ayalew et al. (2004), Lee and Pradhan (2007), Blahut et al. (2010), Yune et al. (2010), and Lee et al. (2012), permits a comparative analysis of the influential attributes and hazard assessment methods of debris flows. Debris-flow hazards are dependent on a specific set of factors and processes that are usually investigated by various experts (such as hydrologists, geologists, or civil engineers). Prediction concerns where and/or when debris flow will occur, and depends on the type of movement and the scope of the forecasting. Many attributes (geography, rainfall, geology, vegetation, wildfire history, and conditions of existing structures) are related to the mechanism/initiation of debris flows (Table 1).

Table 1 Debris-flow influencing factors and existing survey methods

Influencing factor	Influencing attribute	Survey method
Topographical property	Elevation	Digital map
	Slope (angle, direction, and shape)	DEM (digital elevation model)
	Valley (length, slope, and width)	Remote sensing

Influencing factor	**Influencing attribute**	**Survey method**
	Watershed (area and slope)	Satellite image
	Initial zone (area, volume, shape, and failure type)	Field survey
Hydrological property	Rainfall (Max., intensity, recurrence period, and accumulated rainfall)	AWS (automatic weather station)
Geological property	Lithology Bedding plane angle	Geological map
	Moisture content	Field survey
	Void ratio	
	Porosity	
	Saturation	
	Density	
	Permeability	
	Groundwater	
Vegetation property	Tree (species and tensile strength of the root)	Stock map
	Wildfire history	Field survey
Structural	Sedimentation volume	Design file

Influencing factor	Influencing attribute	Survey method
fragility	Size of drainage	
	Fragility curve	

To obtain information on the influencing factors other than topographical properties (elevation, slope, valley shape, and watershed size) and rainfall data, field surveys should be thoroughly conducted throughout entire expressway facility sections. Because the assessment of debris-flow hazards in this study is to be applied on a regional scale for management of the entire national expressway system, the method needs to be simple and applicable to the macrozonation of debris-flow hazards. Other factors, such as the size and shape of a valley along with the variations in slope direction, properties of the subsoil, geological properties, and vegetation cover also have an influence on the movement of debris flows. To simplify the method, only slope information was considered for debris-flow movement possibilities. Because obtaining and processing all the attributes stated above in the prediction stage is a time-consuming and difficult task, we decided that only easily accessible document data (such as digital maps, geological maps, and so on) were to be used in the assessment process. For example, the spatial correlations (for site-specific microscale areas) between influence of debris flow and geological or vegetation characteristics were not well investigated in Korea. Since most debris flows regardless of scale are induced by regional torrential rains, available and reliable influencing attribute data such as rainfall and digital map is necessary to immediately evaluate the debris-flow hazard for a wide coverage area. Consequently, the KEC debris-flow hazard assessment method has become a fundamental assessment tool owing to its simplicity.

As a precautionary measure, the KEC method assesses the hazard of debris flows at a regional scale using a limited number of datasets. Only digital elevation models (DEMs) and expressway design files of the target area were used. The DEMs that were used in the assessment process were those provided by the National Geographic Information Institute (NGII) of Korea, whereas the expressway design files were provided by the Korea Expressway Corporation. Through the application and use of only easily accessible datasets, the KEC hazard assessment method minimizes the need for time-consuming field investigation, and allows rapid hazard assessment of debris flows in large area (administrative sections or nationwide). Accordingly, the DEM for elevation and slope was used to evaluate the watershed information containing topographical route condition of rainfall-induced debris flow. Two indices—the susceptibility value and vulnerability value—were used to evaluate debris-flow hazard.

The susceptibility value indicates the likelihood of debris-flow occurrence in a target area and is assessed using a total of four attributes. The mean watershed slope and area percentage of watershed with slopes over 35° are used to assess debris-flow initiation. Mean valley slope and length percentage of valley with slopes over 15° are used for the assessment of debris-flow movement. Other factors such as the size and shape of the valley along with variations in slope direction, properties of the subsoil, and vegetation also have an influence on the initiation and movement of debris flows. To simplify the method, only the slope information derived from the DEMs is considered for debris-flow possibilities. Each of the influencing factors are assigned points from 0 to 5 based on the grading standard set (Table 2) considering past debris-flow occurrence cases for 10 years (from 2000 to 2009), when the annual precipitation rate amounted to 3000 mm and adds up to a maximum total susceptibility value of 20 points. For the weight considerations of the four attributes, logistic regression was carried out through the Statistical Package for Social Science. The occurrence of debris flow is "one" when it happened and "zero" when there was no occurrence. Logistic regression has been used frequently for multivariable analysis (Ohlmacher and Davis 2003; Ayalew and Yamagishi 2005; Lee and Pradhan 2007). In this study logistic regression is used to estimate the correlation between the dependent variable (occurrence of debris flow) and independent variables (influencing attributes) by evaluating probabilities using a logistic function. Results of logistic regression analyses using datasets from 30 debris-flow cases show that the four susceptibility value attributes had weights of 0.27, 0.24, 0.26, and 0.23, respectively. Since the weights showed no significant difference, the attributes were considered to have identical weights.

The vulnerability value indicates whether a debris flow will actually damage or have an impact on expressway sections. The vulnerability value is assessed by two attributes: the volume of margin area available in which to deposit debris-flow materials before reaching expressway structures; and the size of drainage facilities running through the expressway. For the acquisition of the attribute values, expressway design files provided by the Korea Expressway Corporation were used. Each of the attributes is given points ranging from 0 to 5 based on a grading standard (Table 2), and these points are added up to provide the total vulnerability value of 10 points.

Table 2: Points given to attributes according to the grading standard of KEC

			Scoring Index	Points
Susceptibility Value (20 points)	Initiation assessment (10 points)	Mean slope of watershed (Unit: °)	Higher than 35°	5
			30°–35°	4
			25°–30°	3
			20°–25°	2
			15°–20°	1
			Under 15°	0
		Area percentage of watershed with slopes over 35° (Unit: %)	Higher than 40%	5
			30%–40%	4
			20%–30%	3
			10%–30%	2
			1%–10%	1
			Under 1%	0
	Movement assessment (10 points)	Mean valley slope (Unit: °)	Higher than 25°	5
			20°–25°	4
			15°–20°	3
			10°–15°	2
			5°–10°	1
			Under 5°	0
		Length percentage of valley with slopes over 15° (Unit: %)	Higher than 90%	5
			70%–90%	4
			50%–70%	3
			30%–50%	2
			10%–30%	1
			Under 10%	0
Vulnerability Value (10 points)	Debris storage (5 points)	Volume of deposit area (Unit: m^3)	No accumulation area (0 m^3)	5
			0 m^3–100 m^3	4
			100 m^3–1000 m^3	3
			1000 m^3–5000 m^3	2
			Higher than 5000 m^3	1
			Excessive volume of deposit area, No damage guaranteed	0
	Debris passage through expressway facilities (5 points)	Size of drainage facility (Unit: m)	Waterway	5
			Lateral drains below D1.2	4
			Waterway box below B1.0x1.0	3
			Waterway box below B4.0x4.0	2
			Waterway box exceeding B4.0x4.0 – Discharge section under 30 m^2	1
			Small bridges	0

Source Korea Expressway Corporation (2009)

Note D is the diameter of lateral drains and B is the width of waterway box for the scoring index of debris passage through expressway facilities. Debris passage through expressway facilities is defined as the designed sedimentation volume of debris flow from valley to waterway under the expressway

By combining the calculated susceptibility value and vulnerability value (Fig. 1), a hazard class is given for a target expressway section. In the table of severity rating shown in the chart of hazard class (Fig. 1), the x-axis and y-axis indicate the vulnerability value and susceptibility value, respectively. Through investigations on past debris-flow occurrences, the hazard classes were categorized according to the rainfall recurrence period for expressway design purposes. Hazard class S indicates a likelihood of debris-flow occurrences in areas with rainfall recurrence periods of 2–5 years. Hazard classes A, B, C, and D have rainfall recurrence periods of 5–20, 20–50, 50–100 years, and over

100 years, respectively. Hazard class E indicates an area with a very low likelihood of debris-flow damage (Korea Expressway Corporation 2009).

3 GIS-BASED FRAMEWORK FOR DEBRIS-FLOW ASSESSMENT

Based on the processing of attributes included in the KEC method, a real-time debris-flow hazard assessment framework for expressways was newly developed using the ArcGIS 10.2 software (Fig. 1). Various ArcGIS tools, such as the Spatial Analyst tools and the Analysis tools, were used for a quantitative and objective assessment of the attributes using rainfall datasets on a real-time basis.

3.1 Attribute Processing for the Susceptibility Value

In order to process watershed slope and valley slope datasets, DEMs provided by the NGII of Korea were used. Numerical maps with the highest resolution were those of a scale of 1:1000. But 1:1000 scale numerical maps were only provided for major urban areas. Because numerical maps of the highest resolution provided for the entire Korean Peninsula were those of 1:5000 scale, numerical maps with a scale of 1:5000 were employed in attribute processing to obtain the susceptibility value.

Of the entities within the DEMs, only the polyline and point entities having elevation value were extracted from the numerical map and used for construction of the DEMs. Because the system focuses on the debris-flow hazard assessment of expressway facilities, the expressway layers were selected. For the processing of slopes in the surrounding area of expressways, the elevation layers were also selected (Fig. 2a). With the elevation layers of the DEMs, the elevation and slope raster images with the smallest cell sizes possible were obtained, as shown in Fig. 2b, c. Because the minimum cell size that could be considered with 1:5000 DEMs was 5 m, the raster with cell sizes of 5 m by 5 m was processed. Based on the elevation raster, the flow direction datasets were computed. The Flow Direction tool creates a raster of flow direction from each cell to its steepest downslope neighbor (Olivera et al. 2002a). From the flow direction raster, the flow accumulation datasets were obtained (Fig. 2d). The Flow Accumulation tool creates a raster of accumulated flow into each cell. With a flow accumulation grid as the accumulated weight of all cells flowing into each downslope cell, valleys can be defined through the use of the flow accumulation value (Olivera et al. 2002b) (Fig. 2e). For a more accurate visualization of valley areas, the properties of the flow accumulation grids were altered in various ways. Through trial and error and by comparing with the actual site survey (or the valley layer of the digital map) and the flow accumulation grids, the accumulation grids (white lines in Fig. 2e) having 0.1 STD of valley shape (geometry of valley based on the DEMs) were applied to define the boundary of valleys.

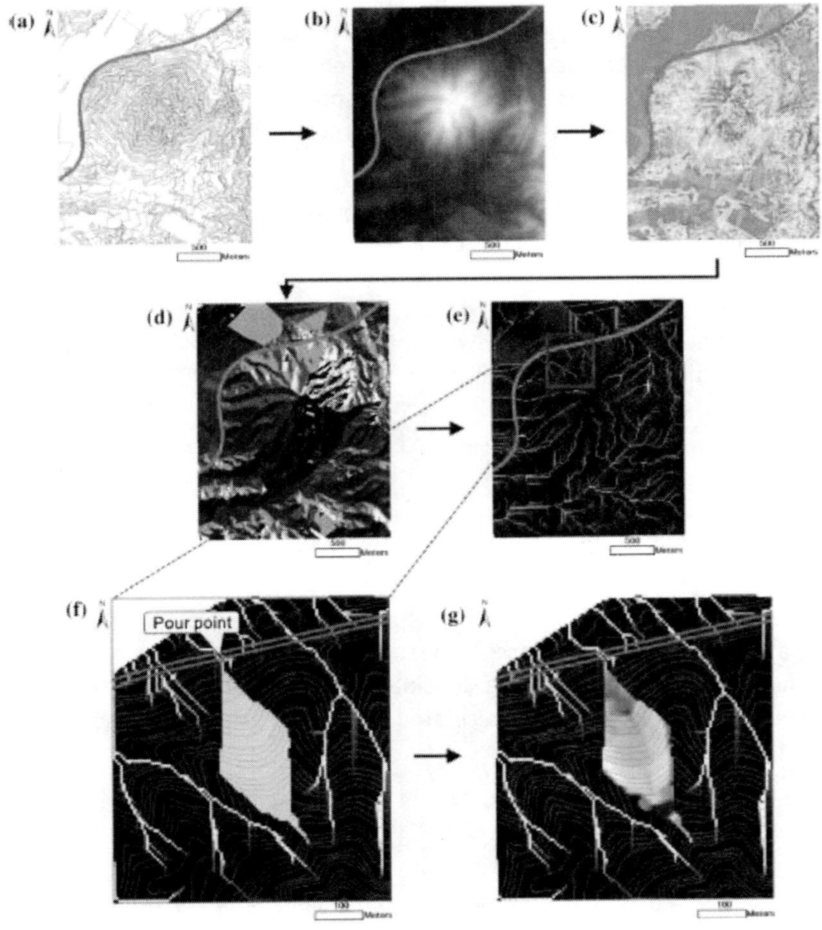

Fig. 2: Details of sequence application (using the proposed GIS-based framework) for watershed slope attributes: **a** polyline and point entity of DEM, **b** DEM for elevation, **c** DEM for slope, **d** flow direction, **e** flow accumulation, **f** watershed area, **g** DEM for watershed slope

After setting a pour point (output point) on the route of the assessed expressway, the flow direction and pour point were taken into consideration to obtain the watershed. The Watershed tool creates the drainage areas that contribute to the flow from the land surface to the water system (Esri 2002) (Fig. 2f). Through the Extract by Mask tool, the slopes of the cell in the watershed area were obtained (Fig. 2g). Through the histogram in the raster properties, the values for the attributes of the mean watershed slope and area percentage of the watershed with slopes over 35° were acquired. Because the slope raster can indicate the steepest slope with regard to the surrounding pixels and the flow direction, but not the slope in the valley direction, the previously obtained slope raster cannot be used to derive the attributes related to the whole valley. Other means should be used for the acquisition of the valley slope data. An approach that assesses the slope through the elevation raster in the valley direction was proposed and applied.

From the flow accumulation layer, the valley shapes in the watershed were obtained. The valley paths were manually plotted on the elevation layer. The elevations of the cells in the path of the valley were obtained through the Extract by Mask tool of ArcGIS using the plotted valley path and elevation layer. With the length of the plotted valley path, and the total elevation difference between the expressway and the highest point in the valley path, the mean valley slope was simply calculated as follows:

$$\text{Mean valley slope } (\Phi_i) = \tan^{-1}\left(\frac{\Delta H}{L_{\text{path}}}\right) \tag{1}$$

ΔH is the elevation difference between the expressway and the highest point in valley path, while L_{path} is the length of the plotted valley.

In the estimation and prediction stage, the valley in which a debris flow will initiate is unknown. Therefore, for watersheds with more than one valley, all valleys in the watershed were considered in the process. The mean valley slopes for each valley were calculated, and were averaged using the valley lengths as weight factors:

$$\text{Overall mean valley slope} = \frac{\sum_{i=1}^{n}(\phi_i \cdot L_i)}{\sum_{i=1}^{n} L_i} \tag{2}$$

Φ_i is the mean valley slope of the individual valley and L_i is the length of the individual valley separated by the watershed DEMs.

The extracted DEMs obtained through the Extract by Mask tool did not properly represent the valley directions, showing abrupt valley direction changes at right angles (Fig. 3a). In order to appropriately assess the slopes in the actual flow directions of the valleys for the calculation of the length percentage of the valley with slopes over 15°, additional ArcGIS tools were used. The extracted valley path elevations of cells were converted to points through the Raster to Point tool, and were assigned to the map coordinate system (Fig. 3b). With the Buffer tool, a buffer of 5 m was set around the manually plotted valley path (Fig. 3c). In order to obtain the valley path points that are positioned inside the 5 m buffer zones of the plotted valley path, the Intersect tool was used. Through the process, only the valley path points that were in the vicinity of the actual valley path were obtained (Fig. 3d). Using the shape value (as file type of ArcGIS) of extracted points, the slope between the DEMs in the direction of the valley travel path was calculated using Microsoft Excel. Through the calculation results of the type of shape files, the distances between cells with slopes of over 15° were obtained, which in turn allowed the calculation of the length percentage of a valley with slopes over 15°. According to the mean valley slope calculation process in watersheds with more than one valley, the length percentage of valley with slopes over 15° was calculated using a similar process:

Overall length percentage of valley with slopes

$$= \frac{\sum_{i=1}^{n}\{(L_{15})_i\}}{\sum_{i=1}^{n} L_i} \quad (3)$$

$(L_{15})_i$ is the length of valley with slopes over 15° for the individual valley.

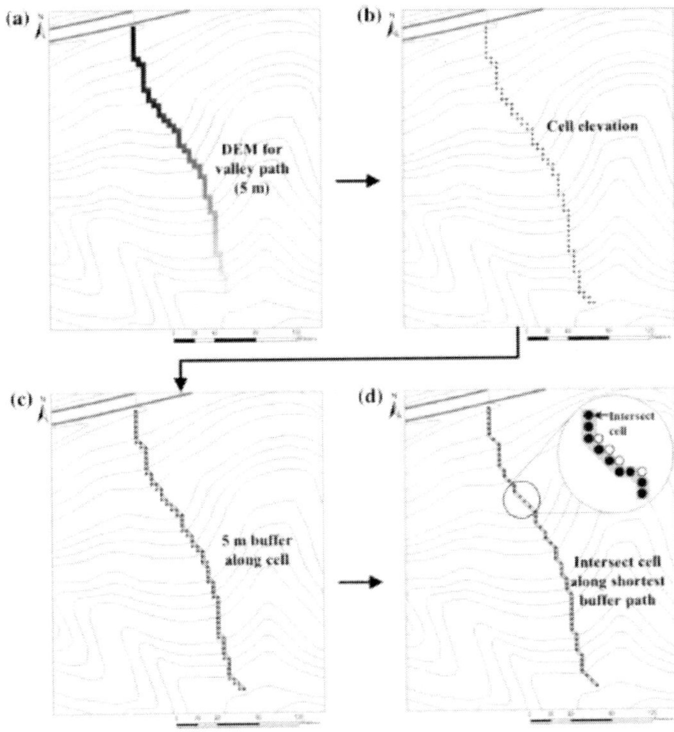

Fig. 3: Details of sequence application (using the proposed GIS-based framework) for valley slope attributes: **a** elevation extracted for cells on valley path, **b** cell elevation converted to points, **c** buffer applied on valley path, **d** intersect of buffer and cell points

3.2 Attribute Processing for the Vulnerability Value

From the expressway design files (Korea Expressway Corporation 2009), the volume of area available for sedimentation was calculated by simplifying the area as a triangular pyramid bounded by the valley and expressway embankment (Fig. 4). The volume of the margin area of sedimentation is calculated using Eq. 4. The drainage facility information of each watershed was obtained through the expressway design files, and the attribute points were determined according to the grading standard set by the KEC (Table 2).

$$\text{Sedimentation volume } (\mathrm{m}^3) = \frac{1}{6}(L \times H \times W) \tag{4}$$

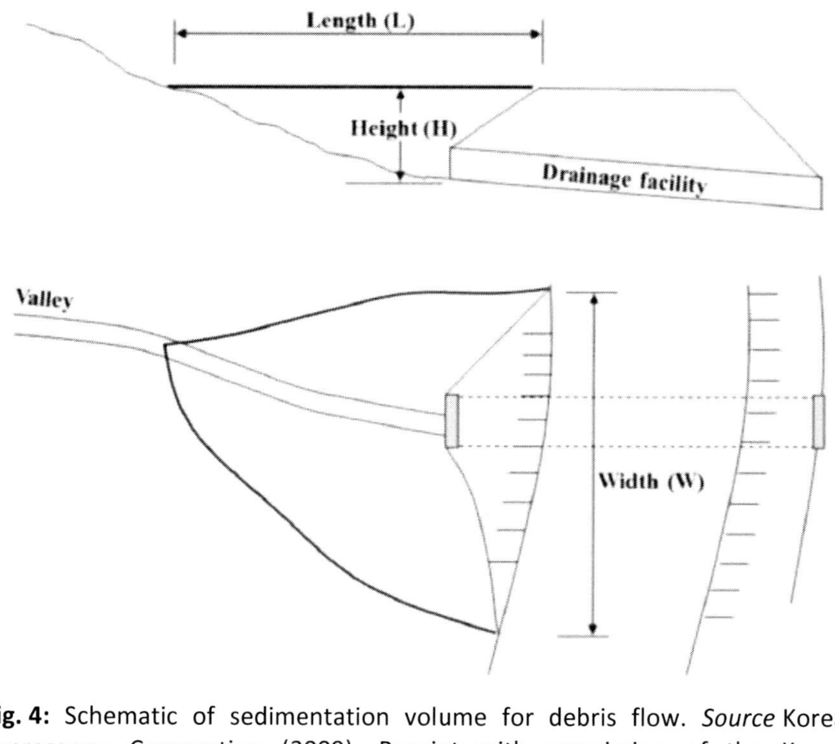

Fig. 4: Schematic of sedimentation volume for debris flow. *Source* Korea Expressway Corporation (2009). Reprint with permission of the Korea Expressway Corporation

3.3 Quantification of Rainfall Criteria for Hazard Class

According to the KEC method, the rainfall recurrence periods (ranging from 2 years to more than 500 years) were used as trigger values of the landslide hazard. But to immediately determine the rainfall event by linking the network sever of meteorological observatories and real-time estimates of the landslide hazard, it is impossible to directly apply the rainfall recurrence periods for the proposed framework. Thus, rainfall recurrence periods were quantified as representative accumulated rainfall criteria: 1 h rainfall (Yune et al. 2010), 6 h rainfall (Oh and Park 2013; Ham and Hwang 2014), and 3 day rainfall (Yoo et al. 2012). The baseline rainfalls were set at the lower bounds of 1, 6 h, and 3 day rainfall data from AWS and at the major interchanges at 18 stations on the nine expressway sections in which the 163 debris-flow events (the highest number of debris-flow events from 2000 to 2009) occurred in Korea. Also, various sets of these rainfall baselines using accumulated rainfall data recorded from AWS were cross-validated with the actual damage state for early warning. These referred to the recurrence intervals of rainfall corresponding to the hazard class in the KEC-based debris-flow assessment (Choi et al. 2015). From the correlations between recurrence period and accumulated rainfall data regarding to the hazard class, the range and lower bound of 1, 6 h, and 3 day accumulated rainfall datasets were determined (Table 3). Thus, the rainfall criteria were set with the lower bound values of accumulated rainfall ranges for five hazard classes based on conservative hazard management.

Table 3: Quantified rainfall criteria for hazard class based on 163 debris-flow events with accumulated rainfall records

Hazard class	**Rainfall recurrence period (year)**	**1 hours rainfall (mm)**		**6 hours rainfall (mm)**		**3 days rainfall (mm)**	
		Accumulated rainfall range	**Rainfall criteria**	**Accumulated rainfall range**	**Rainfall criteria**	**Accumulated rainfall range**	**Rainfall criteria**
S	2–5	34.5–47.7	35	88.6–122.8	90	218.7–317.7	220
A	5–20	47.7–62.2	45	122.8–167.2	125	317.7–443.7	320
B	20–50	62.2–71.4	60	167.2–195.1	160	443.7–527.4	420
C	50–100	71.4–78.3	75	195.1–216.3	195	527.4–586.8	520

Hazard class	Rainfall recurrence period (year)	1 hours rainfall (mm)		6 hours rainfall (mm)		3 days rainfall (mm)	
		Accumulated rainfall range	Rainfall criteria	Accumulated rainfall range	Rainfall criteria	Accumulated rainfall range	Rainfall criteria
D	Higher than 500	Higher than 94.2	95	Higher than 265.0	270	Higher than 724.5	720

The severity levels of debris flow were defined as three levels: safe, caution, and warning (Fig. 1). If the measured rainfall datasets do not correspond to all three cases of accumulated rainfall datasets, the target expressway section is considered as safe from debris-flow hazard. If more than one indicator exceeds the three baseline rainfalls with regard to the five hazard classes, the debris-flow hazard needs to be forecast as a level of caution or warning. The criteria are appropriate for conservative risk management because of the cross validation between rainfall datasets and hazard class in the case of actual damage status on the representative nine expressway sections (see Sect. 5.2). The quantified rainfall criteria for the hazard class of debris flow create difficulties for real-time management of debris-flow hazard on the national expressway, although it is desirable to apply the probability of the possible rainfall recurrence period (Choi et al. 2015). Therefore, simulations using relationships between recurrence period and accumulated rainfall data at locations where debris flows have occurred is required to establish the appropriate rainfall criteria for debris-flow risk assessment.

4 SAFETY MANAGEMENT SYSTEM FOR DEBRIS-FLOW HAZARD

The GIS-based safety management system for debris-flow hazard consists of a database (DB) and four systematic modules (Fig. 5). The database contains all field data and processed data in the system. The submodules execute specific functions that manage and utilize information in the database. These include the input module, the debris-flow hazard assessment module, the decision-making module of debris-flow countermeasures, and the real-time, debris-flow hazard assessment module linked to the rainfall monitoring network.

The framework, which includes all these functions, focuses on user-friendliness and real-time applications.

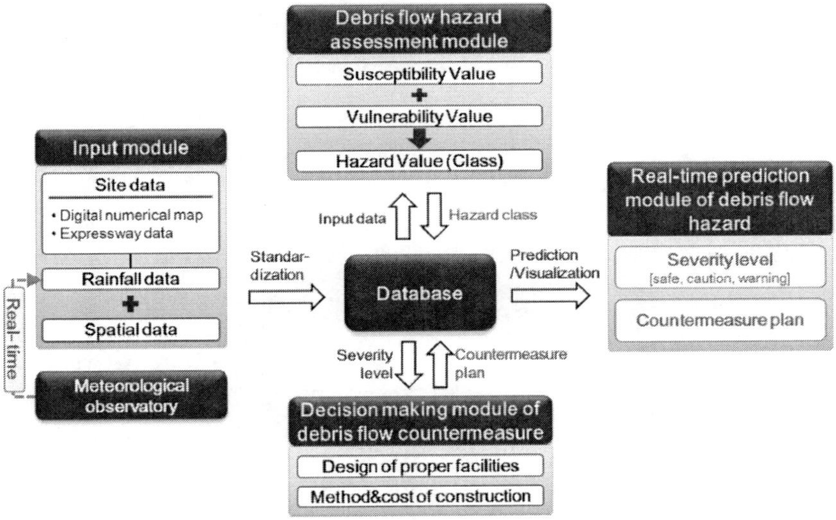

Fig. 5: Integrated framework with debris-flow hazard assessment procedures

4.1 System Database

A database is the backbone of the developed debris-flow framework. It stores not only collected data such as the digital numerical map, expressway, and real-time based rainfall monitoring data (which is standardized using AWS, transmitted from meteorological observatory server), but also contains processed data obtained from the debris-flow hazard assessment and real-time prediction modules (Fig. 6). The DB contains all data as alphanumeric values according to standard data formats, which represent the outcome of data classification and standardization with spatial information. The data stored in the database can be easily utilized in the framework. The digital numerical map can be used as the basic topographical information of the system because it offers an easy way to construct topographical information for a target area. The expressway information is composed of organization categories for managing Korean expressway route data, and spatial datasets designed for various coordinates systems (longitude and latitude, GRS80, Google coordinates). The data format of rainfall events (belonging to the AWS network) was arranged previously for an expressway in Korea. The general information about rainfall observatory and time-series rainfall monitoring information were standardized considering real-time connection with AWS network. Furthermore, the results of hazard analysis and alert information were built into the database at the same time as the framework of debris-flow assessment.

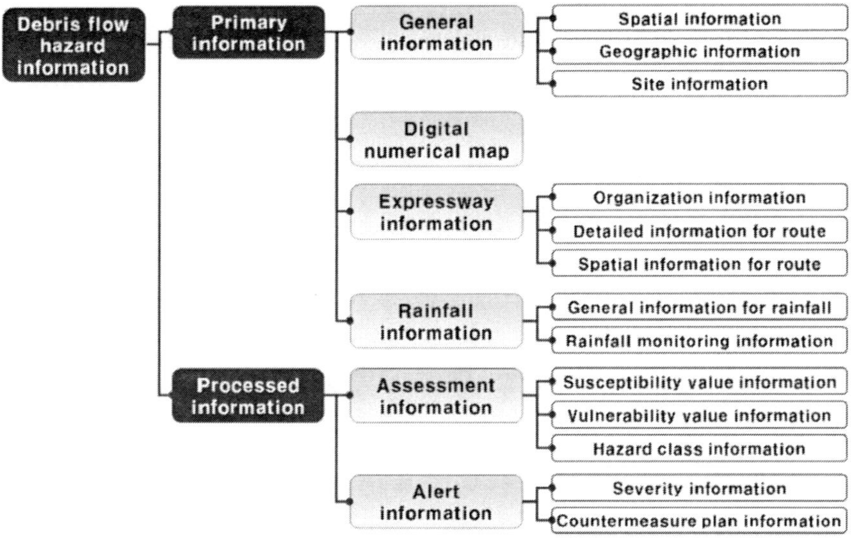

Fig. 6: Key data classes and relations of the database for the developed system

4.2 Submodules for Rainfall and Safety Management

Input function provides an effective way to store and arrange all collected data, including electric or nonelectric documents, general information, digital numerical maps, expressway data, rainfall monitoring data, and analyzed data, according to a standardized data format in the DB. Before applying the rainfall criteria for severity levels of debris flows, the target expressway sections are grouped within the spatial range of influence of AWS observatory (5 km radius), around the nearest AWS in Korea (Fig. 7). The influencing areas vary in different localities of a watershed, but we group the nearest expressway sections in accordance with the average distance (5 km). In order to link the site-specific rainfall datasets with expressway sections, additional investigation about influencing areas of AWS observatories considering the hydrological characteristics of mountain areas is important, but this study did not consider this complexity. The reclassified AWS datasets for rainfall criteria (1, 6 h, and 3 day accumulated rainfall) are linked to the grouped expressway sections using the system database.

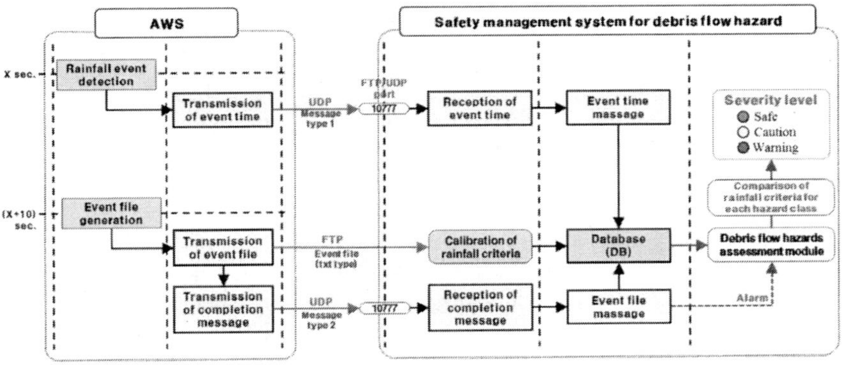

Fig. 7: Schematic flow of transmission protocol of rainfall datasets from AWS to safety management system for debris-flow hazard

Using the real-time-based rainfall transmission module, as the rainfall accumulation records are transmitted from the AWS in real time, the rainfall event protocol is retained to consider the event trigger time and transferred event file. In Fig. 7, as soon as a rainfall event is detected, the rainfall event time is transmitted from the AWS to the developed system. Then, the generated event file (stored in a oracle server) is transferred in company with the completion message, which is the computational message to inform the transmission or reception of event file and enable the DB, to operate the safety management system. At the developed system server, an event file is calibrated based on the rainfall criteria, according to the file transfer protocol. The rainfall accumulation datasets, including completion message of event time and file, are stored in system database. The alerted rainfall datasets are compared with rainfall criteria for hazard class (Table 3), and the severity level can be determined in 30 s after the occurrence of the rainfall event.

The real-time debris-flow assessment framework has four functional phases within the database based on the proposed schematic sequence of the debris-flow hazard assessment module (Fig. 7). In the first phase, linked with the digital numerical map and the DEM, the watershed DEM and valley layer are extracted using the ArcGIS desktop program and input system DB. In the second phase, the susceptibility value and vulnerability value for the target route are constructed into DB and combined with geospatial information. In the third phase, reliable rainfall monitoring data for the target route is transmitted from the widely distributed AWS server on a real-time basis, the routes of the completed site investigation for debris-flow hazards are grouped into the same datasets and focused on the adjacent rainfall station in target areas. The rainfall values for debris-flow hazard assessment are automatically computed based on the monitoring criteria with reference to recurrence periods of rainfall (Korea Expressway Corporation 2009 when the rainfall monitoring data is input to the DB. Once the rainfall threshold level for debris-flow hazard is reached (from the KEC method), the severity levels (safe, caution, and warning) are determined using map symbols (blue, yellow, and red symbols, respectively, in Fig. 7) at the target route in real time. In addition, sound signals and message windows are alarmed to notify the expressway administrator of the hazard status.

A decision-making module of debris-flow countermeasures was developed by standardization of site investigation reports as part of a nation-wide expressway survey in mountainous areas of Korea. As soon as the severity levels based on the rainfall monitoring data in real time are decided, it is possible that optimum countermeasures for debris-flow minimization can be rapidly put in place for a threatened target route depending on whether a caution or warning severity level is activated. Facilities (ring net, debris barrier, and so on) for debris-flow prevention can be preposition beforehand based on the site investigation for target routes having potential of debris flow. The method and cost of construction can be studied and visualized in a report, corresponding to the anticipated severity level in near real time.

The debris-flow hazard prediction function shows all the attributive information in the database by using tables and graphics according to its characteristics, on screen or as a document. Also, all data in the DB can be exported as a chart or graph. The graphic functions simultaneously display interpolated data and field data over a designated area. All of the charts and graphs can then be printed. Especially, the debris-flow hazard can be visualized and forecast as 2D maps overlaid with satellite images. Also, the severity level of the target route can be determined using zonation criteria in real time.

In this proposed framework, the computer-based method for the real-time assessment of spatial debris-flow hazard was embedded based on a stand-alone system developed using Microsoft Visual BASIC and the Esri ArcGIS developer tool (Lee and Wong 2001; Esri 2006). The ArcGIS developer tool was mainly used for the development of the database, evaluation of the results, and spatial visualization. Several assumptions and precedent assessments are needed to estimate the possible debris-flow hazard for a target site in real time, at the point at which the debris flow occurs. Especially, the precedent procedures consist of the building of the database, DEM construction, and determination of susceptibility value and vulnerability value in order to consider the overall debris flow potential in the target area, prior to the occurrence of rainfall and debris flow. As the debris flow occurs near the target site, the possible severity level can be estimated in real-time by using the rainfall data monitored from the AWS server.

5 APPLICATION OF THE ASSESSMENT METHOD ON SELECTED SITES

The developed real-time GIS-based debris-flow hazard assessment framework is applied to three debris flow-damaged expressway areas in Korea to verify its validity. The standard set of the original KEC method was subsequently revised based on the case study.

5.1 Condition of Application

The Pyeongchang area of the Yeongdong Expressway, the Deogyu Mountain area of the Daejeon-Jinju Expressway, and the Juksan-Geochang area of the 88 Expressway in Korea were selected for investigation of their debris-flow hazards (Fig. 8). Many debris flows occurred in the Pyeongchang area of the Yeongdong Expressway during a heavy rainfall event (244.0 mm/day, 66.0 mm/h) in the summer of 2006 (Table 4). Debris flows occurred in the Deogyu Mountain area of the Daejeon-Jinju Expressway in the summer of 2005. The rainfall intensity in the region at that time was 312.0 mm/day and 54.5 mm/h. The Juksan-Geochang area faced several debris flows during a heavy rainfall event in the summer of 2006. The rainfall intensity at that time was 121.0 mm/day, and 31.5 mm/h. When comparing all target areas, the Pyeongchang and Deogyu Mountain areas had similar rainfall intensities, indicating the same rainfall recurrence periods, whereas the Juksan-Geochang area had the lowest daily rainfall. For each area, a test bed area was set for the application of the assessment method. A test bed with a length of 11 km along the Yeongdong Expressway was chosen for the Pyeongchang area. The lengths of the test beds for the Daejeon-Jinju Expressway and 88 Expressway were 3 and 2 km respectively, presented as black dotted lines in Fig. 8.

5.2 Verification of the Framework for Debris-Flow Assessment

All existing watersheds in the selected expressway test beds were analyzed. Of all the watersheds in the selected regions, the areas with expressways positioned on bridges and tunnels, or near vast areas of fields were excluded from the analysis due to their very low likelihood of damage. Since reported debris flows were based on the damage made to road structures, those areas without any damage were not reported. No debris-flow damage was reported in areas where the space for possible sedimentation deposit is vast, and consequently these areas were not considered in the analysis process. After the exclusion of low-risk sites, watersheds were then classified according to whether or not debris-flow damage had been reported. Areas with debris-flow damages reports are referred to as "occurrences," and those without damage reports are referred to as "nonoccurrences." As a result, 18 debris-flow occurrences and 14 nonoccurrences were analyzed for the Pyeongchang area (divided into 32 subsections) based on the proposed framework for debris-flow assessment; 12 debris-flow occurrences and 8 nonoccurrences were analyzed for the Deogyu Mountain area (divided into 20 subsections); and 8 debris-flow occurrences (evaluated as 1 warning and 7 caution level according to the developed system) and 8 nonoccurrences were analyzed for the Juksan-Geochang area (divided into 16 subsections) as shown in Fig. 9. The developed safety management system for debris-flow hazard constructed the debris-flow hazard map according to hazard class for watershed and forecasted severity level linked with AWS (hourly max rainfall datasets) on real-time basis at the Jusan-Geochang area (Fig. 9).

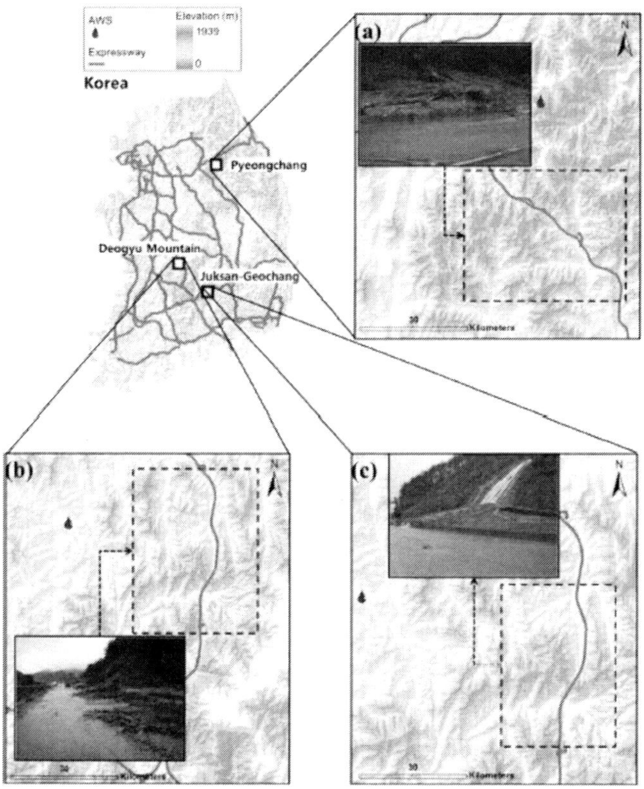

Fig. 8: Overview of assessed target regions and pictures of damaged area: **a** Pyeongchang area of the Yeongdong Expressway, **b** Deogyu Mountain area of the Daejeon-Jinju Expressway, **c** Juksan-Geochang area of the 88 Expressway. *Source* Korea Expressway Corporation 2009). Reprint with permission of the Korea Expressway Corporation

Table 4: State of debris-flow hazard and measured rainfall of three target expressway sections

Expressway section	Time of occurrence	Rainfall (mm)		Damage of debris flow	
		Daily accumulated rainfall	Hourly max rainfall	Condition	Debris sedimentation (m^3)
Pyeongchang area	15 July 2006, 12:00	244.0	66.0	Road blocked	5000
Deogyu Mountain area	3 August 2005, 01:00	312.0	54.5	Road blocked	3000
Juksan-Geochang area	18 July 2006, 23:00	121.0	31.5	Drainage blocked	1500

Applications of the KEC method show results that roughly coincide with the actual debris-flow occurrences and nonoccurrences are presented in the table of hazard value (varying from 0 to 30) and hazard class (Fig. 10). Occurrence cases are roughly positioned in the upper right-hand side, which indicate higher susceptibility and vulnerability values, whereas nonoccurrence cases are located on the lower left side, with relatively lower susceptibility and vulnerability values. Although this tendency may seem correct to some extent, it does not always show flawless results. In the C and D hazard classes, both debris-flow occurrence and nonoccurrence cases are mixed up, not always indicating a result in which occurrences have higher hazard classes, and nonoccurrences fall into lower classes. Debris flows occurred even in areas with an E hazard class, which was initially expected to indicate a very low likelihood of debris flow. The total sum of susceptibility value and vulnerability value even showed a greater number of nonoccurrences than the occurrences due to the low hazard values of occurrence cases.

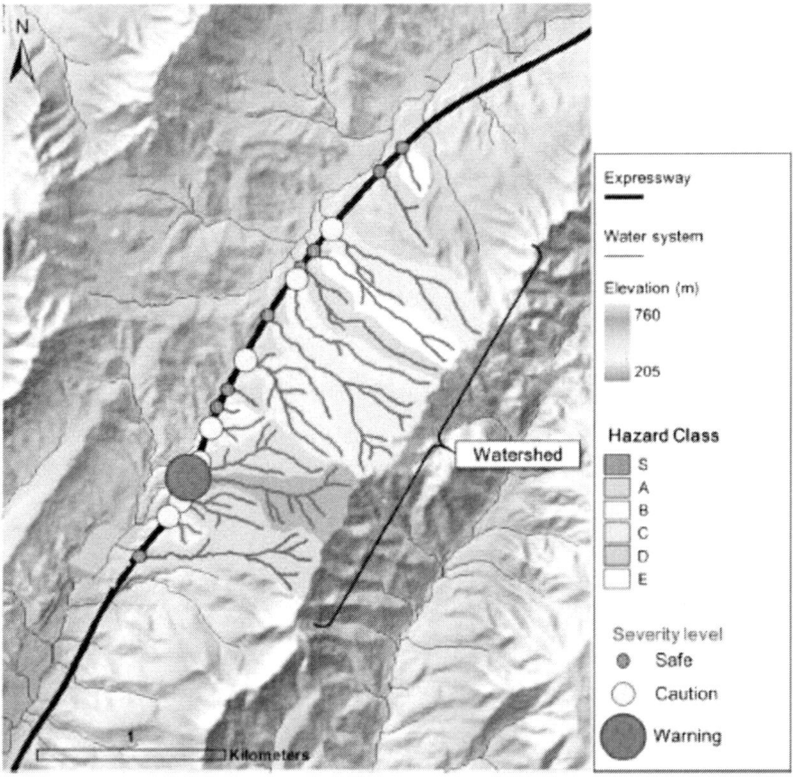

Fig. 9: Example of debris-flow hazard map based on the hazard class for watershed and severity level by simulation through real-time link with AWS in the Jusan-Geochang area at 23:03, 18 July 2006. At that time, the daily accumulated rainfall and the hourly maximum rainfall were recorded as 121.0 and 31.5 mm, respectively

Although the sites showed results that generally presented higher susceptibility values and vulnerability values in occurrence cases, they did not perfectly represent the total hazard value differences. But the effectiveness of the proposed framework was validated by the three cases because the potential of debris-flow hazard is generally high for the actual occurrence cases. The average of the hazard value of occurrence cases is 1.13 times that of nonoccurrence cases (Table 5). In order to appropriately represent the differences between occurrences and nonoccurrences of debris flows, modifications can be made on the grading standard of the KEC method using more validation tests based on the proposed framework, because the KEC standard set was established using logistic regression with a small number of samples of debris-flow damaged expressway sections. Specifically, grading standards for attributes need to be revised by taking into consideration the attribute values of both occurrences and nonoccurrences, maximizing the distinction between the two instances.

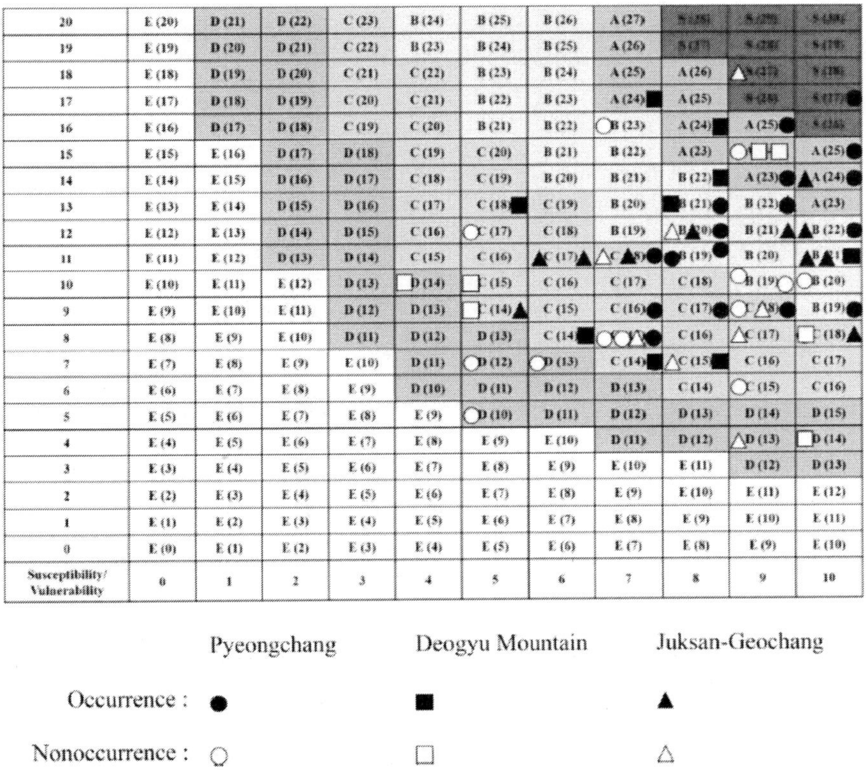

Fig. 10: Original chart of hazard class in accordance with susceptibility and vulnerability values of debris flow and occurrence and nonoccurrence cases for three expressway sections

5.3 Modifications to the Framework

The existing KEC method assigned points to each attribute according to a grading standard set based on a few case studies of debris-flow occurrences. But it did not take the nonoccurrence cases into consideration (Kim 2012). In addition, the KEC grading standard set used all data samples collected for the study. In order to appropriately represent the differences between debris-flow occurrences and nonoccurrences, modifications can be made to the grading standard using additional tests. The three applications in this study included a large number of debris-flow occurrence sites and can be used for this purpose.

Table 5: Average of susceptibility value, vulnerability value, and hazard value of occurrence and nonoccurrence cases according to the original standard-based verification results

Test beds	Occurrence case			Nonoccurrence case		
	Susceptibility value	**Vulnerability value**	**Hazard value**	**Susceptibility value**	**Vulnerability value**	**Hazard value**
Pyeongchang	11.67	8.56	20.22	9.36	7.64	17.00
Deogyu Mountain	11.78	7.44	19.22	10.14	7.44	17.57
Juksan-Geochang	11.25	8.33	19.58	9.63	8.25	17.88
Average	11.57	8.11	19.67	9.71	7.78	17.48

To modify the grading standards, the average values of attributes in the three target areas were calculated. When calculating these average values, the mismatched cases (between estimated severity level and actual damage) were eliminated from the calculation. Likewise, occurrence cases that showed considerably high hazard values, such as 25 and higher, and nonoccurrence cases with total values of 11 and lower, were considered as outliers for statistical normalization and were not considered when calculating the average values. After calculating the average values of each attribute for the occurrence and nonoccurrence cases of the three target areas, the highest attribute values of the occurrence and nonoccurrence cases were respectively defined as the upper and lower boundaries of each scoring index. The range of each scoring index of the four susceptibility value attributes was divided into six classes using equal intervals. Logistic regression was performed to reestablish the grades of scoring indices in the range of 0–5. Results of logistic regression analysis for the four susceptibility value attributes indicate that the six grades have weights of 0.29, 0.24, 0.19, 0.14, 0.10, and 0.05, respectively. Accordingly, the redefined grades of the scoring indices were assigned points from 0 to 5. The modified susceptibility grading standard is presented in Table 6. The vulnerability grading standard was also modified based on the debris-flow occurrence and nonoccurrence cases of the three target areas (Table 6).

Table 6: Modified points given to attributes based on the original grading standard of KEC

			Scoring Index	Points
Susceptibility Value (20 points)	Initiation assessment (10 points)	Mean slope of watershed (Unit: °)	Higher than 30°	5
			30°–28°	4
			28°–26°	3
			26°–24°	2
			24°–22°	1
			Under 22°	0
		Area percentage of watershed with slopes over 35° (Unit: %)	Higher than 32%	5
			32%–25%	4
			25%–18%	3
			18%–11%	2
			11%–4%	1
			Under 4%	0
	Movement assessment (10 points)	Mean valley slope (Unit: °)	Higher than 19°	5
			19°–17°	4
			17°–15°	3
			15°–13°	2
			13°–11°	1
			Under 11°	0
		Length percentage of valley with slopes over 15° (Unit: %)	Higher than 67%	5
			67%–61%	4
			61%–55%	3
			55%–49%	2
			49%–43%	1
			Under 43%	0
Vulnerability Value (10 points)	Debris storage (5 points)	Volume of deposit area (Unit: m^3)	No accumulation area (0 m^3)	5
			0 m^3–100 m^3	4
			100 m^3–500 m^3	3
			500 m^3–2000 m^3	2
			Higher than 2000 m^3	1
			Excessive volume of deposit area, No damage guaranteed	0
	Debris passage through expressway facilities (5 points)	Size of drainage facility (Unit: m)	Waterway	5
			Lateral drains below D1.2 (1.31 m^2)	4
			Waterway box below B2.0x2.0 (4 m^2)	3
			Waterway box below B4.0x4.0 (16 m^2)	2
			Waterway box below B3.0xD3.0 (26 m^2)	1
			Bridges	0

Note D is the diameter of lateral drains and B is the width of waterway box for the scoring index of debris passage through expressway facilities. Debris passage through expressway facilities is defined as the designed sedimentation volume of debris flow from valley to waterway under the expressway

Although the grading standard of the original KEC method was established based on 30 field applications (KEC 2009), the scale of damage by debris flow during the considered period was very small (Kim 2014). Considering the fact that only a very small number of the considered sites had deposit areas with volumes exceeding 2000 m^3, the grading standard was modified to reflect this range, that is, the highest class of debris storage was altered from above 5000 m^3 to above 2000 m^3. The other standards were also modified accordingly.

Based on the same condition of application for the three target areas, the hazard value and hazard class were determined using the revised grading standard. On the scoring chart, the occurrence cases are distributed in hazard classes S, A, and B, which have a high potential of debris flow, as shown in Fig. 11. In contrast, the points of nonoccurrence cases are distributed in hazard class C, D, and E, which have comparatively lower potential for debris flow. In addition, the average of the hazard value of occurrence cases is 1.58 times that of nonoccurrence cases (Table 7); this is significantly higher than the result (1.13) using the original KEC standard. Moreover, the method has potential to be improved with more data accumulation and more case histories. Also attributes other than those regarding slope should be considered, such as watershed size and valley curvature. According to Kim (2014), with larger watershed sizes both the debris-flow initiation risk and occurrence risk increase. With larger valley bending ratios, more debris-flow materials are subjected to sedimentation, lowering the possibility of damage on road structures.

Fig. 11: Modified chart of hazard class in accordance with susceptibility and vulnerability values of debris flow applying the modified scoring criteria and occurrence and nonoccurrence cases for three expressway sections

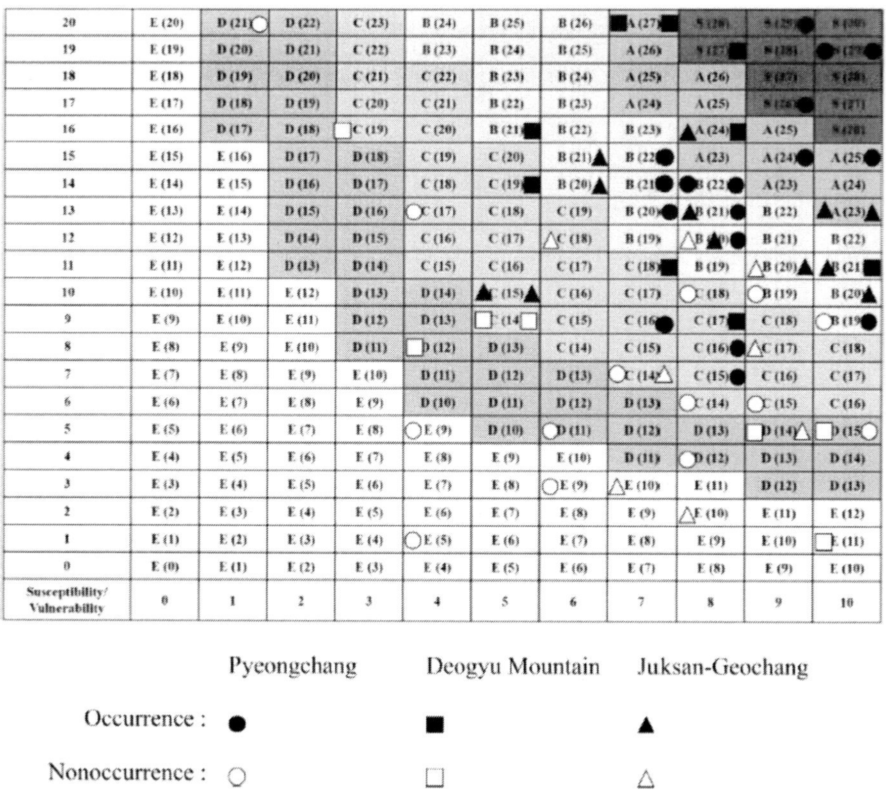

Table 7: Average of susceptibility value, vulnerability value, and hazard value of occurrence and nonoccurrence cases according to the modified standard-based verification results

Test beds	Occurrence case			Nonoccurrence case		
	Susceptibility value	**Vulnerability value**	**Hazard value**	**Susceptibility value**	**Vulnerability value**	**Hazard value**
Pyeongchang	13.53	9.50	23.03	8.52	5.62	17.00
Deogyu Mountain	14.62	8.52	23.14	9.01	6.92	17.57
Juksan-Geochang	15.31	9.02	24.33	7.12	7.55	17.88
Average	14.47	9.01	23.50	8.22	6.70	14.91

6 CONCLUSIONS

In this study, to provide preliminary information for rapid decision making of evacuations or restoration measures, and to prevent secondary disasters caused by debris flows, a GIS-based real-time framework of debris-flow hazard assessment for expressway sections was proposed based on the KEC debris-flow hazard assessment method. First, the KEC-based method was standardized using Esri ArcGIS, which enabled various attribute datasets to be acquired in an objective and quantitative manner with a fixed data acquisition sequence. Second, considering the real-time link with the AWS network, the rainfall recurrence periods were quantified as accumulated rainfall using trigger values of actual debris-flow events. Based on the optimized KEC debris-flow hazard assessment method, a safety management system for debris-flow hazards was developed based on the GIS platform. For field applicability evaluation, the framework was applied on three expressway sections in Korea: the Pyeongchang area of the Yeongdong Expressway, the Deogyu Mountain area of the Daejeon-Jinju Expressway, and the Juksan-Geochang area of the 88 Expressway. The reliability of the assessment method was investigated by comparing with actual debris-flow occurrence and nonoccurrence cases. A revised grading standard set was proposed and validated in a case study of damaged expressways in comparison with the KEC method. In the future, additional attributes including watershed size and bending of valley should be considered in the hazard assessment.

Footnotes

1 www.kma.go.kr/weather/observation/aws_table_popup.jsp.

ACKNOWLEDGMENTS

This research was supported by the Basic Research Project of the Korea Institute of Geoscience and Mineral Resources (KIGAM) and the National Research Foundation of Korea (NRF) Grant (No. 2015R1A5A7037372) funded by the Korean Government (MSIP). The authors sincerely acknowledge the support and guidance provided by the editors and reviewers. We especially thank the Korea Expressway Corporation for its leadership and support.

REFERENCES

1. Ayalew, L., H. Yamagishi, and N. Ugawa. 2004. Landslide susceptibility mapping using GIS-based weighted linear combination, the case in Tsugawa area of Agano River, Niigata Prefecture, Japan. *Landslides* 1(1): 73–81.
2. Ayalew, L., and H. Yamagishi. 2005. The application of GIS-based logistic regression for landslide susceptibility mapping in the Kakuda-Yahiko Mountains, Central Japan. *Geomorphology* 65(1): 15–31.
3. Blahut, J., P. Horton, S. Sterlacchini, and M. Jaboyedoff. 2010. Debris flow hazard modelling on medium scale: Valtellina di Tirano, Italy. *Natural Hazards and Earth System Sciences* 10(11): 2379–2390.
4. Carrara, A., F. Guzzetti, M. Cardinali, and P. Reichenbach. 1999. Use of GIS technology in the prediction and monitoring of landslide hazard. *Natural Hazards* 20: 117–135.
5. Choi, J.S., J.W. Jeong, O.G. Kwon, C.K. Chung, and S.D. Lee. 2015. Simulation of debris-flow early warning using real time rainfall monitoring. *Proceedings of 2015 KGS Spring National Conference*, 19–20 March 2015, Seoul, Korea, 25–29.

6. Dai, F.C., and C.F. Lee. 2002. Landslide characteristics and slope instability modeling using GIS, Lantau Island, Hong Kong. *Geomorphology* 42(3): 213–228.

7. Esri. 2002. *Arc Hydro*. California: Esri Press.

8. Esri. 2006. *ArcGIS 9: Using ArcGIS Desktop*. California: Esri Press.

9. Feusto, G., C. Alberto, C. Mauro, and R. Paola. 1999. Landslide hazard evaluation: A review of current techniques and their application in a multi-scale study, Central Italy. *Geomorphology* 31: 181–216.

10. Ham, D.H., and S.H. Hwang. 2014. Review of landslide forecast standard suitability by analysing landslide-inducing rainfall. *Journal of Korean Society of Hazard Mitigation* 14(3): 299–310.

11. Hungr, O., S.G. Evans, M.J. Bovis, and J.N. Hutchinson. 2001. A review of the classification of landslides of the flow type. *Environmental & Engineering Geoscience* 7(3): 221–238.

12. Iverson, R.M. 2000. Landslide triggering by rain infiltration. *Water Resources Research* 36(7): 1897–1910.

13. Jakob, M., and O. Hungr. 2005. *Debris-flow hazards and related phenomena*. New York: Springer.

14. Keller, E.A., and D.E. DeVecchio. 2008. *Natural hazards: Earth's processes as hazards, disasters, and catastrophes,* 2nd edn. Upper Saddle River, NJ: Prentice Hall.

15. Kim, K.S. 2012. *Analysis of rainfall characteristics inducing shallow failure of road cut slope*. Ph.D. Dissertation, Seoul National University, Seoul.

16. Kim, S.R., H.S. Kim, G.S. Kim, and C.K. Chung. 2014. Debris-flow risk assessment along expressways in Korea using GIS. *Proceedings of Geohazards 2014*, 20–21 November 2014, Kathmandu, Nepal, 159–164.

17. Korea Expressway Corporation (KEC). Expressway and Transportation Research Institute. 2009. Development of debris flow hazard analysis method and its application. Expressway and Transportation Research Institute Research Report.

18. Kritikos, T., and T. Davies. 2015. Assessment of rainfall-generated shallow landslide/debris-flow susceptibility and runout using a GIS-based approach: Application to western Southern Alps of New Zealand. *Landslides* 12(6): 1051–1075.

19. Lee, J., and D.W.S. Wong. 2001. *Statistical analysis with ArcView GIS*. Toronto, ON: Wiley.

20. Lee, S., and B. Pradhan. 2007. Landslide hazard mapping at Selangor, Malaysia using frequency ratio and logistic regression models. *Landslides* 4: 33–41.

21. Lee, S.W., G.H. Kim, C.Y. Yune, H.J. Ryu, and S.J. Hong. 2012. Development of landslide-risk prediction model through database construction. *Journal of Korea Geotechnical Society* 28(4): 23–33.

22. Lin, P.S., J.Y. Lin, J.C. Hung, and M.D. Yang. 2002. Assessing debris flow hazard in a watershed in Taiwan. *Engineering Geology* 66: 295–313.

23. Oh, J.R., and H.J. Park. 2013. Establishment of landslide rainfall threshold for risk assessment in Gangwon area. *Journal of Korean Society of Hazard Mitigation* 13(3): 43–51.

24. Ohlmacher, G.C., and J.C. Davis. 2003. Using multiple logistic regression and GIS technology to predict landslide hazard in northeast Kansas, USA. *Engineering Geology* 69(3–4): 331–343.

25. Olivera, F., J. Furnans, D. Maidment, D. Djokic, and Z. Ye. 2002. *Drainage systems, ArcHydro: GIS for water resource*. Redlands, CA: ESRI Press.

26. Olivera, F., D. Maidment, and D. Honeycutt. 2002. *Hydro networks, ArcHydro: GIS for water resource*. Redlands, CA: ESRI Press.

27. Park, D.W., N.V. Nikhil, and S.R. Lee. 2013. Landslide and debris flow susceptibility zonation using TRIGRS for the 2011 Seoul landslide event. *Natural Hazards and Earth System Sciences* 13: 2833–2849.

28. Paulín, G.L., M. Bursik, J.L. Hubp, L.M.P. Mejía, and F.A. Quesada. 2014. A GIS method for landslide inventory and susceptibility mapping in the Río El Estado watershed, Pico de Orizaba volcano, México. *Natural hazards* 71(1): 229–241.

29. Wang, C., S. Li, and T. Esaki. 2008. GIS-based two-dimensional numerical simulation of rainfall-induced debris flow. *Natural Hazards and Earth System Sciences* 8: 47–58.

30. Yoo, N.J., D.H. Yoon, J.K. Um, D.G. Kim, and B.S. Park. 2012. Analysis of rainfall characteristics and landslides at the west side area of Gangwon Province. *Journal of the Korean Geo-Environmental Society* 13(9): 75–82.

31. Yune, C.Y., K.J. Jun, K.S. Kim, G.H. Kim, and S.W. Lee. 2010. Analysis of slope hazard-triggering rainfall characteristics in Gangwon Province by database construction. *Journal of the Korean Geo-Environmental Society* 26(10): 27–38.

12

DEVELOPMENT OF METHODS AND PRACTICES OF VIRTUAL REALITY AS A TOOL FOR PARTICIPATORY URBAN PLANNING: A CASE STUDY OF VILNIUS CITY AS AN EXAMPLE FOR IMPROVING ENVIRONMENTAL, SOCIAL AND ENERGY SUSTAINABILITY

Gintaras Stauskis

Vilnius Gediminas Technical University, Sauletekio ave. 11, Vilnius LT-10223, Lithuania

ABSTRACT

Background

Multiple channels of sustainable urban development could be managed in the process of an inter-disciplinary cooperation and creative involvement of all stakeholders. The paper aims to verify how the specific methods of virtual reality simulation could be used as an effective tool for setting an efficient participation platform between the stakeholders to improve environmental, social and energy sustainability of urban development in modern cities. The paper is based on the case study of Vilnius City.

Methods

The article shows how innovative tools could facilitate better cooperation based on more balanced approaches in participatory urban planning by chronological review of traditional public involvement practices and structured analysis of emerging public participation techniques. As each discipline involved in sustainable urban development covers specific socio-economic, environmental, economic and creative competence fields, it is essential to develop a comprehensive tool for public participation that is well accessible, reliable and efficient.

Results

The provided analysis of research theories and practical applications of public involvement through virtual simulation in the case of Vilnius City shows that virtual reality and spatial modelling integrated on gaming platforms under the conditions of widespread information and communications technology (ICT) applications could be developed and used as an innovative, involving and efficient tool for facilitating wider and more effective public involvement in the planning and design of urban environment and so increase social and environmental sustainability of urban development process. The application of this method for planning public spaces leads to more efficient use of environmental resources also reducing the need for primary energy and using it more rationally.

Conclusions

The analysis of using virtual urban simulation (VUS) method for representing, analysing, referring and improving urban design projects reveals that virtual applications can make public participation more accessible for citizens of the community and improve reliability of its results thus having direct impact on the planning process and its outcomes in terms of more sustainable impact on environmental and economic development of urban community.

Keywords: Urban planning, Landscape, Virtual reality, Public participation, Stakeholder involvement

Tell me, I forget,

Show me, I remember,

Involve me, I understand,

Step back, I will act.

Old Chinese proverb

BACKGROUND

Emerging public participation

The public initiatives that emerged in the early 70s in the UK and other European countries are good cases to analyse and illustrate how public participation gradually turned into an instrument of urban planning and design as we understand and use it today. The post-war policy of rebuilding the ruined blocks in big British cities turned into the method of urban renewal when the entire residential areas were pulled down on the reason of low sanitary quality and the goal of rising living standards. In reality, this brought demolition of old development in Birmingham, Manchester, Glasgow and other cities. That process of 'improving' was additionally fuelled by rising income of urban councils as a result of relatively dynamic economic development [1]. Several drawbacks of this urban policy were fast to appear. On the one hand, valuable urban legacy was destroyed; the entire urban communities with specific lifestyles and cosy neighbourhoods were completely erased. On the other hand, the modernist style of concrete blocks that appeared on their former land sites were sold fast but appeared to be disgusting to live in and were abandoned in a short term by their residents turning the whole areas into urban deserts. Last but not least, complete rebuilding was a very costly venture compared to gradual refurbishment of existing stock. The process of protests of local residents was fast to appear, and it was professionally supported by architects (Figure 1). In Manchester, an architect Rod Hackney and his neighbours established a Black Road Area Action Group to protect historic residences from demolition and went into a massively supported campaign of debates with the city council called 'Save my house' in 1972.

Figure 1: **Residents looking for real participation power in urban redevelopment of the UK.** Real people need real power before Megatropolis strikes their home. Images: Communities in Control [2]; http://www.ourpatch.com.au/system/attachments/0003/3949/Megatropolis.jpg.

The success of a particular case in Manchester turned into a precedent for the whole country that led to wider discussions on the issues of urban redevelopment between the residents and the councils. The case of hearing local residents and then setting the refurbishment strategies was finally formalised and included into the UK's housing law. Still, efficient methods of communication between the clerks, the public and the business were to be developed. In Europe, many countries have gone through a similar process like in the UK in approximately the same decade. As a matter of fact, modern public participation methods were not used in East European countries until the last decade of the twentieth century for evident socio-political reasons. Finally, regulations on public involvement found their stake in national construction law mostly following the examples of other West European countries. In general perspective, public participation facilitates decisions on more rational planning and design thus saving material and social resources including urban land, materials, labour and energy.

METHODS

The paper is designed as a set of analysed practices and case studies with public participation in planning. Techniques and potentials of spatial modelling and virtual representation are reviewed and analysed. The research starts by presenting the development of techniques in virtual representation and turning them into a planning tool. Virtual simulation in gaming environment is analysed by presenting several case studies including the one developed by the author of this paper. By comparing different tools and methods used for public participation, their strengths and weaknesses are identified. By comparative analysis, passive, active and interactive public participation methods are structured in the aspect of their interactivity. As a conclusion, the recommendations are drawn on how to apply and replicate innovative information and communications technology (ICT)-based public participation tools for better results in urban development.

Research of virtual reality applications in urbanism

Since the first papers were published on research in virtual reality and space simulation in 1989, the immense list of publications is available, illustrating the scope of research activities ongoing in different knowledge centres around the globe [3]. As noted by Simpson in 2001, the published research primarily focused on general issues of environment simulation later calling for application of virtual reality in planning and design as well as analysing current examples and the role of urban simulation [4]. Along that road, the aspects of urban simulation in geographic information system (GIS) environment were analysed, reflecting state of the art in virtual urban modelling. Applications of simulation for hazard research are one more area for research. Wider development of research on virtual reality applications in urbanism collide with technical limitations of both hardware and software, as well as lack of established research methodology. The most advanced research centres focus on the complex research of three-dimensional city scale simulation in a GIS environment also combining it with dynamic building of integrated modelling systems. Several main topical categories of references in virtual reality simulation could be identified. *Urban and environmental simulation* is a broad area covering references in urban and environmental simulation, visualisation and virtual reality, focusing on contributions to urban planning. *Virtual reality and GIS* application is another field for researchers whose centre of their glance is on virtual reality and urban simulation as it is developed in GIS environment. *Urban and regional modelling* by building, evaluating and testing urban and regional development models is the next broad category of references. *Urban simulation and gaming* section explores simulation and technical aspects of gaming. *Technical aspects* of urban and environmental simulation are common category of research being presented independently or incorporated into one of the above-mentioned categories.

Virtual reality models are different from reality, so it is an impression of a person involved in experiencing virtual reality. There are important aspects of human perception of the presence in virtual models that empower human senses as memory, bodily movement and kinesthesis [5]. The sets of parameters such as awareness, memory, sensory notion, feeling of realism, attention and interaction are offered to identify the notion of mixed presence of the user in virtual models.

Stakeholder involvement as planning method

Special regulation adopted in the UK for developing a statement of community involvement (SCI) in urban planning allows the local government to set a goal in endorsing this regulation, to fulfil its vision in order to engage the people and organisations of Sheffield in planning for sustainable development [6]. The experienced benefits of exercising this practice in Sheffield, the third largest metropolitan district in the UK, promote planning as a proactive tool for delivery of sustainable development, reflection of needs of local community, quality improvement and efficiency of planning as well as promotion of social cohesion by giving communities a tangible steak in decision making. The involvement covering public, private, community and voluntary sectors of local urban settlements has to be transparent, promoting real participation and involvement, accessible and inclusive, also accountable. The principles of SCI stated hereby include culture of engagement, early and continuous involvement, reaching out and fit for purpose, clarity and formal representation based on a certain code of practice. Several participation tools are identified to be applied by SCI: media, letters, public meetings, websites, distributed documents, exhibitions and questionnaires. Continuing the involvement, using its results and giving feedback on implementation are most important aspects for local communities.

The stages of SCI are evolving in the process of its undertaking as more cooperation experience is accumulated by the stakeholders. Starting with informing local residents about the planning process that is going to happen, it moves to the next phase: consultation of local residents and survey of their opinions, and further on, it leads to active involvement of communities in becoming a stakeholder and partner of the planning process [6]. The necessity to consider the needs and expectations of all citizens' groups is meant as 'equality' , an aspect which is important for the residents of all groups to believe that the participation process is legitimate and timely organised where the results are used for the public interest [7].

As underlined by the International Association of Public Participation in their project aimed at the improvement of cross-cultural exploration of public governance and decision making, innovative use of communication technology can improve the quality of public participation with indigenous communities who have much better perception of simulated images compared to two-dimensional maps and drafts [8].

However, several decades of applying traditional tools to involve the public into the planning turned out to be restricted in terms of access and time, transparency of the process as well as needed resources. Different types of meetings, committees and working groups spend much time as participants have little practice in understanding planning strategies, the process itself and its outcomes. Additional time is needed to access all groups of citizens, especially the teenagers and the youth. The meetings involving the members of local communities are usually dominated by few most active members who reveal their own opinion more than that of the whole community. It happens that these 'activists' have some motivated interest in the outcome of planning in terms of market competition, and so transparency becomes an issue difficult to control. Multiple meetings, repeated access attempts and the need to control the process involve professional staff, premises and services; therefore, the process usually becomes long and expensive compared to its results.

Several of the above-mentioned methods are practiced in Lithuania as regulated by the Guidelines for Public Participation in Urban Planning [9]. Public participation is outlined therein as 'Voluntary participation of legal and private entities and their groups in preparation of territorial planning document'. Providing information about the goals of planning on the Internet and in local paper, placing it at the local authority's premises, making an exhibition of the project and holding public hearings are the main activities required by the code. Completed planning projects are registered by the local authority at the publicly accessible Planning Register. Time frames for different stages of informing and involving the public depend on a scale of planning: for general plans, that in Lithuania's case covers big urban areas, and all spatial, social and environmental aspects' time intervals are longer; for special plans, covering some specific planning aspects, time intervals are relatively shorter (Table 1).

Table 1: Basic phases of public participation as regulated by the Planning Law of Lithuania

Type of plan/time span for procedures	**General plan**	**Special plan**
Information for citizens about planning procedures	Immediately	Immediately
	Letters to owners, residents, companies	Letters to owners of neighbouring and planned plots
Getting acquainted with plan	1 month	20 working days
Public exhibition time	15 working days	10 working days
Public meeting	10 working days in advance	10 working days in advance

Type of plan/time span for procedures	General plan	Special plan
Complains accepted from public	1 month	1 month

The practice of implementing participatory planning methods in Lithuania has demonstrated quite similar limitations as in the other countries, transparency and efficiency being the main ones. There were cases when organised groups of community representatives have severely criticised selected development projects but later supported similar ideas presented by another developer. Citizen's groups often act against everything and anything without a constructive dialogue, and for this reason, the architects in Lithuania are very modest developing more interactive dialogue and tend to have just formal procedures usually limited to a purely bureaucratic paperwork of announcements, protocols and reports. In this case, the real voice of residents remains unheard.

Accountability of planning organisers and transparency of planning process are the main social gains that are generated in the process of public participation as noted by the EU Directive on Public Participation [10]. Involvement of groups of residents and other NGOs is especially encouraged as it brings also environmental education for people. Wider use of electronic media is appreciated by the Directive as one of the means for appropriate information and involvement.

The basic methods used for public participation based on their interactivity are structured as passive, active and interactive (see Table 2). Passive methods are based mainly on one-way communication between the planners and the public as one of the stakeholders with no interactivity. Active methods have some elements of interactivity as they try to establish 'a dialogue' instead of 'a monologue'. The emerging ICT-based techniques are attributed to interactive methods as their goal provides a person or representative group with possibilities of direct access and reliable and monitored feedback on the goals, roles, process and consequences of planning.

Present challenges in urban development and limitations of traditional participation methods in Vilnius City were the reasons of looking for new more efficient methods of public participation in urban planning and design. Numerous cases popped out in recent years in Vilnius City when centrally prepared development programmes and projects met severe opposition from urban residents: demolition of old housing units at Šnipiškės area close to modern skyscrapers (Figure 2a); protest of Pilaitė housing district residents against the municipality's plan to urbanise recreation area at a local lakeshore (Figure 2b). These protests left local authorities confused and city planners disappointed. On the other hand, attempts of planners, architects and authorities to establish a constructive dialogue with the residents practically failed because of lack of experience from both sides as well as the proposed methods showed up as inefficient and non-transparent.

Table 2: Structure of public participation methods in urban planning and design

Method	Passive	Active	Interactive
Preparatory	Data collection	Field study, Interviews	Internet, websites
Analytic	Case study analysis, survey	Working groups	Policy dialogue
Disseminative	Printed material	Presentation, visualisation	Open space technology, gaming and multiplayer
Monitoring	Observation	Committees, meetings, test audit	E-government, virtual urban simulation

a b

Figure 2: **Resident's initiative and public meeting. (a)** Resident's initiative - a dashboard at Šnipiškės neighbourhood, 2012. Image: http://laimikis.lt/snipiskieciai-kaimynu-lenta-fino-gatve/. **(b)** Public meeting of Pilaitės local community, Vilnius, Lithuania, 2011. Image: http://www.pilaitesbendruomene.lt/istorija.html.

Analysis of the analysed cases shows that the attempts to maintain a contact with local residents and involve them in planning were carried out in primitive ways by using just formally required methods with no follow-up and real impact on the planned development projects. There were several cases during the last years when certain urban development programmes and decisions of city bureaucrats that were not supported by a part of Vilnius citizens ended with social protest of local communities. On the other hand, just the real cases when citizen's rights to make impact on city development are violated cause active informal actions during which the civil society builds up and real participation experience is developed. Also, this encourages local authorities who act as planning organiser to seek for more efficient participation tools to transparently involve all citizens' groups into a dialogue with planners and thus increase planning quality and make positive impact on the city development.

Some cases show how Internet databases used as virtual knowledge platforms with good access have attracted big number of community and professional users who generate their feedback referring to the presented planning and design information in the way of direct comments on website also that brings learning and replicating the provided practices in a wider scale. Sustainable energy innovations that were applied at Grand Spa Lietuva hotel and spa complex in Druskininkai town in Lithuania (Figure 3a) were presented on the sustainable building platform Construction21 [11]. In relatively short time, that attracted lots of virtual visitors nationwide who analysed the provided information and left the feedback. What is more, by reviewing information on the newly refurbished spa complex in the famous Lithuanian resort, town citizens learned about the benefits of innovative smart grid system for residential areas. This triggered interest in replicating such practices and increased visibility of professionals who have designed, installed and are maintaining these facilities. All this was made possible by providing virtual access to urban development cases for the community users on the Green Building platform Construction21. It turned out that the provided information is easier to access and better to understand when presented in a virtual way with simulated architectural images and supplemented by technical data.

Innovative public involvement systems also cover energy and climate issues that are the major challenges in modern communities especially in creating urban complexes of nearly zero-energy buildings (Figure 3a). In the above-mentioned case, with quite limited interactivity, the citizens demonstrated their interest in the newly built up facility, it could be expected that even greater interest would emerge when the future development would be presented in the form of virtual immersive models with interactive user's interface. This approach is illustrated below.

a

Figure 3: Grand Spa Lietuva hotel and detailed urban model. (a) Grand Spa Lietuva hotel, conference centre and spa complex in Druskininkai, Lithuania. Green Building platform: http://www.construction21.eu. Pictures by G. Stauskis. **(b)** Detailed urban model with automated vehicles and pedestrians [12].

Virtual reality is more and more often used as a planning tool. Visualisation of planning and design material gives an important planning support in landscape architecture design. Virtual reality models allow the citizens to understand spatial and temporal processes of design; they encourage communication and comprehend planning proposals. Testing the virtual reality (VR) models in three-dimensional (3D) spatial design showed that 2D models and aerial photos are also needed to support orientation on-site, and the best results are achieved by combining these tools together. The virtual models offered to the public are preferred to be realistic but not too photo-realistic as fancy details and other visual effects distract attention instead of focusing on essential features of a site. Visualisation sparks public interest in planning proposals, while a balance of live VR shows and Internet visits should be maintained. The development of the planning proposal in time is of a great interest both to the public and professionals which is already a 4D task for future development that enables to show and see 3D spatial planning that changes in a particular time frame (Figure 3b) [12].

Possible applications of virtual urban simulation include projects with public involvement into presenting, analysing and evaluating urban development models. Interactive tools provide residents with possibility to get a close look at the projects to understand the concept and feedback their opinions. The virtual urban simulation (VUS) system in association with GIS involves more data about the area and the residents such as land plot data, availability of government redevelopment grants and resident's characteristics including willingness to participate in area redevelopment process. By linking VUS and GIS, it is easy and comfortable for the user to access all types of required information by highlighting it or identifying in other ways. On the other hand, it is doubtful that availability of the mentioned data layers and accessing it in real time will automatically generate information about the suitability of particular land parcels for redevelopment [12].

Effective visual representation of designed models gets especial importance in landscape architecture. It was experienced by the researchers of interactive landscape planning project in Koenigsluter am Ulm (Germany) that the integration of innovative media tools with participation methods in planning process allows achieving better citizen's understanding and higher participation rates [13]. The platform comprised open-source components turned to be efficient for individual implementation depending on the specifics of the project and real needs of the community.

Some space simulation systems enable the user to modify the urban development model: change houses, trees and other elements from the designed library. The option of change of the area in time is especially a useful feature for local community residents, e.g., experiencing the change of landscape in time how trees grow up and streetscape changes over time. Linking these applications to real estate information as land plot costs makes them especially useful. Some systems build their efficiency on features as generating development of the area by selecting houses through a typical prototype library as practiced in Los Angeles, CA, USA.

SimCity video game series which started in 1989 is recognised as one of the first virtual urbanism gaming applications that besides its gaming aspects bears a didactic role that is being used in several urban training practices around the world. Since spatial modelling tools were developed, they adhere to similar features of gaming products, e.g. graphical features, user interfaces, multi-level interaction and others. Some agent-based modelling features are presented as conceptual gaming tools [14]. Video games are more and more important as spatial modelling tools and simulation engines able to analyse socio-spatial and hierarchical systems of modern urban environment both in real and designed cases. Still, video games and simulation models implement rules based on different and even opposite methods: complex models explicitly expose possible choices and their interrelations, while in video games, the user has to discover the rules of the play himself. On the other hand, video games are appealing and pleasant because of their graphic features and interactive scenarios that are more involving and attractive for the user.

Virtual simulation and gaming in urban context

Being a way of learning about the world in general and learning new knowledge in particular, gaming has several great advantages of applying virtual simulation tools for public interaction and, in particular, for participation reasons. Firstly, it is an immediate report of behavioural and emotional status of a participating subject to the system. That allows recording extremely reliable information about the user's opinion without distortion of results for the reasons of delay in time, location, emotional status, different context and other casual circumstances. Second, it is attractiveness and involving a sense of gaming that erases the barriers of age, gender, social experience and others. Thirdly, ICT application provides equal and easy access to the information regardless of time and location.

Virtual simulation is used to facilitate stakeholder involvement in the projects of very different nature. While establishing a multi-stakeholder platform for sustainable urban mobility, the method of simulation is envisaged as one of the important tools for public involvement into presenting, evaluating and even designing urban mobility systems [15]. They are used in order to better understand dynamic interactions of communities within and with each other, as an integral part of involving community into strategic development of the Chicago North Illinois Planning Commission to achieve citizen's participation in many innovative ways. The network of stationary and mobile devices is involved into polling and analysing mobility planning results by using a 'Paint the Town' programme where residents could designate areas for future development or redevelopment, visualise future land uses and track its impact on transportation, housing, economy, environment and other sectors of the city life (Figure 4). All the presented methods are used to turn usual PC gaming applications from entertaining leisure activity to a creative tool that - in our article's case - is used for the tasks of urban design and architecture.

ICT-based gaming tools used to specifically involve youth in the urban planning process activate the specific segment of citizens that are often ignorant of urban policies as long as they stay aside of the youngster's life. The application of Minecraft Block by Block as an ICT tool for young people can show planners how they would like to see their city's public spaces, residential and recreation areas look like in the future [16]. The partnership of the UN-Habitat, Swedish Building Services and the gaming company aims to give the youth greater appreciation of how sustainable urban development could be integrated into their cities.

Figure 4:**Community members using 'Paint the Town' technology to design and track changes in mobility of Chicago, IL, USA, 2013.** Image: http://www.internationaltransportforum.org/IntOrg/ecmt/urban/Washington03/Thomas.pdf.

Virtual urban simulator for public participation

The methodology of public participation presented hereby has been developed at the Department of Urban Design at Vilnius Gediminas Technical University (VGTU), Lithuania in partnership with Vilnius City Municipality's Urban Development Department. ICT and virtual gaming support is provided by the researchers of the Department of Graphical Systems at VGTU. As a case study, VUS method is applied for verification of preliminary design of the Missionaries public park in Vilnius City (Figure 5). It combines 3D modelling and virtual space simulation with community involvement techniques and provides a tool for analysing and transcribing the results in a usable form.

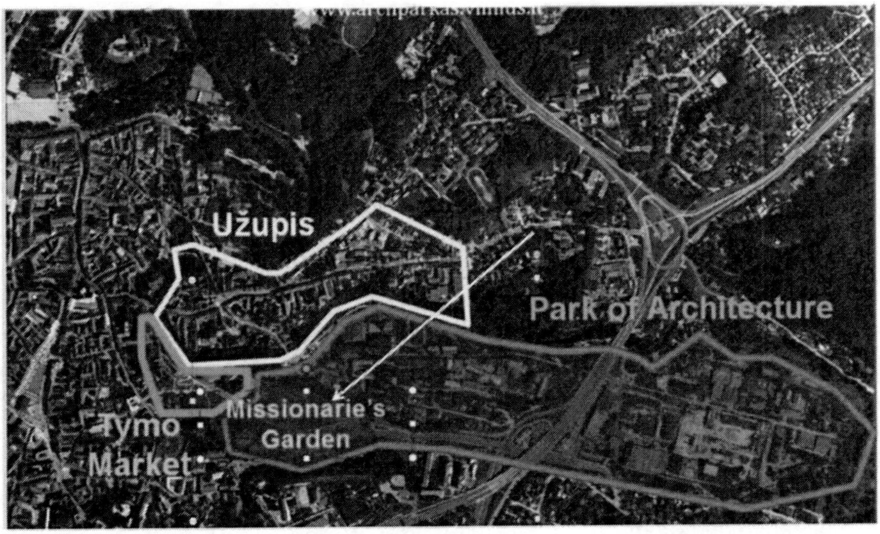

Figure 5: Location map of the VUS case study - Missionaries Park in Vilnius City. Property of Vilnius City Urban Development Department.

The basic problem of designing Missionaries Park is its better social inclusiveness and response to the needs of local communities and all citizens. Municipal authorities that are responsible for the project are not sure if the draft programme for the park design is fully coherent with these needs and wishes to verify it with the citizens. The draft project of the park regeneration has been prepared and presented to professional community where it was evaluated in the aspect of professional quality. Still, social inclusiveness is the goal pursued by the local authority also for the reason to make construction and maintenance of the park easier by involving local business and public interest. The Missionaries site is located on the edge of Vilnius Old Town, at the site of the historic suburb. In fact, it is a central part of town within the historic residential area on one side and an abandoned industry facility on the other. The area is a part of the former Missionaries convent established on-site in the sixteenth century (Figure 5). Proper regeneration of the park with integrated services from the surrounding buildings is expected to bring one more recreational facility within the historic part of Vilnius City.

There were several basic features set as an initial aim of developing the new system for public participation in planning and design. Primarily, it is the immediate character of response that would allow eliminating uncertainties with participant's real opinion regardless of external influences. Secondly, it is possible to maintain access to the focus group of residents who, in our understanding, are interested in planning outcomes and impacts on their community. Thirdly, it is easily accessible for any user with a PC, laptop, tablet or even a mobile phone device used by the young and the elder, at home, office, public space or elsewhere. What is needed more is permanent access to the planning data on different stages of a project. From preliminary design to development plan, implementation, monitoring and feedback are main stages on which the efficiency of the system has to be designed, tested and evaluated.

The important step of the VUS project is setting the criteria for evaluating the plan. In the VUS case, the project team has developed two basic groups of aspects to be surveyed: efficiency of social environment in terms of space use, building integration and services, and the other group of criteria - visual emotional impact of environment with quality of space perception, visual impact of plants and quality of virtual representation (Figure 6). Both groups of criteria are equally important as they characterise the overall impression and opinion of the user at personal and community levels in functional and compositional aspects. The detailed list of criteria is developed to reflect on the social and visual aspects of the analysed park (Table 3). The concrete qualitative and quantitative information accumulated in the process of gaming is analysed and evaluated under all these criteria to provide the project group with the summary report on the four main aspects for the survey that are raised by the organiser and the client of this study - Vilnius City Municipality: how does the Missionaries Park would work for the needs of all Vilnius citizens, how would it satisfy the needs of the local community, is the Park well shaped to host different types of events, and finally is it a safe place for recreation (Table 3)?

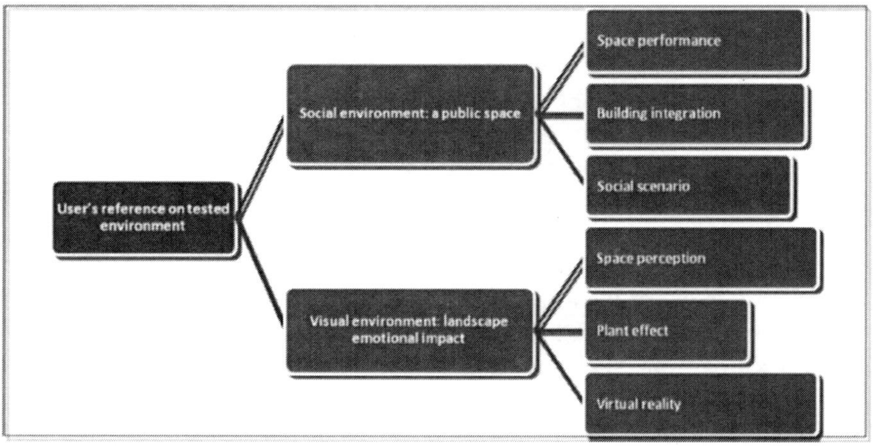

Figure 6" Criteria chart for analysing user's involvement in virtual urban simulation. By G. Stauskis.

Table 3: Survey indicators for analysing virtual gaming information

Survey questions	Indicators	Score
a. Can the designed Missionaries Park function as *the central recreation space of the city*?	• Park is frequently used by all urban residents	6
	• Park is the popular meeting and communication place	6
	• Park with services from surrounding buildings is the family leisure place	6
	• Spatial arrangement and plant design facilitate public communication in park	5
b. Can Missionaries Park become the part of life *for local communities*?	• Local residents daily visit the park	6
	• Residents have their loved places and spaces in park	6
	• Local residents organise their events in the Park	6
	• Local residents take part in planning, construction and maintenance of the park	5
c. Is Missionaries Park capable to attract mass *cultural and recreation events*?	• Poetry, music and other festivals take place in the park	6
	• Park spaces and plants attract wedding and other event photography	6
	• Flower and other exhibitions and competitions take place in the Park	5
	• Park spaces host cultural events of the neighbouring convent	5

Survey questions	Indicators	Score
d. Can Missionaries Park become the *safe and secure place for recreation?*	• All spaces in the park are attended by different groups of users	10
	• Spatial arrangement of the park makes it free from areas for potential attack and vandalism	10
e. Other questions for the survey	• Park entrances and fencing	3
	• Pedestrian transit tracks across and in the park	2
	• User-safe water features	2
	• Features and services of surrounding buildings are frequently used	2
	• Most popular viewpoints and visual connections to and from the park	3
	Total score	100

In order to build an efficient communication between Vilnius municipality as the client, community as the user and the research as the service provider, the VUS methodology was developed in the stages of preparation, involvement and analysis (Figure 7). The preparation phase included task assignment and selection of the project site, building up a 3D model (Figure 8) and providing it with gamification features with necessary functions and characters as movement and collision physics, interaction with environment elements, introduction of human characters and other features. During this phase, the municipality as the client gave the research group the task to check the preliminary plan and learn the importance of the designed park for the city and for the local community, to check its attractiveness for organising different events and also to evaluate its safety and security for its users. Every sub-phase is important to the overall success of the project application. Selecting a proper intake area gives the number of population to be involved into the process, and in the VUS case, it was limited to 10 min reach from the neighbouring urban blocks what is appropriate for the dense old town development with several other recreation areas in similar time distances. It might be different in, e.g. new residential environment where distances measured in travel time are greater. In the next phases of the project, the method of Space Syntax will be used to determine the proper intake area according to the typology of urban grid of the planned and neighbouring areas [17].

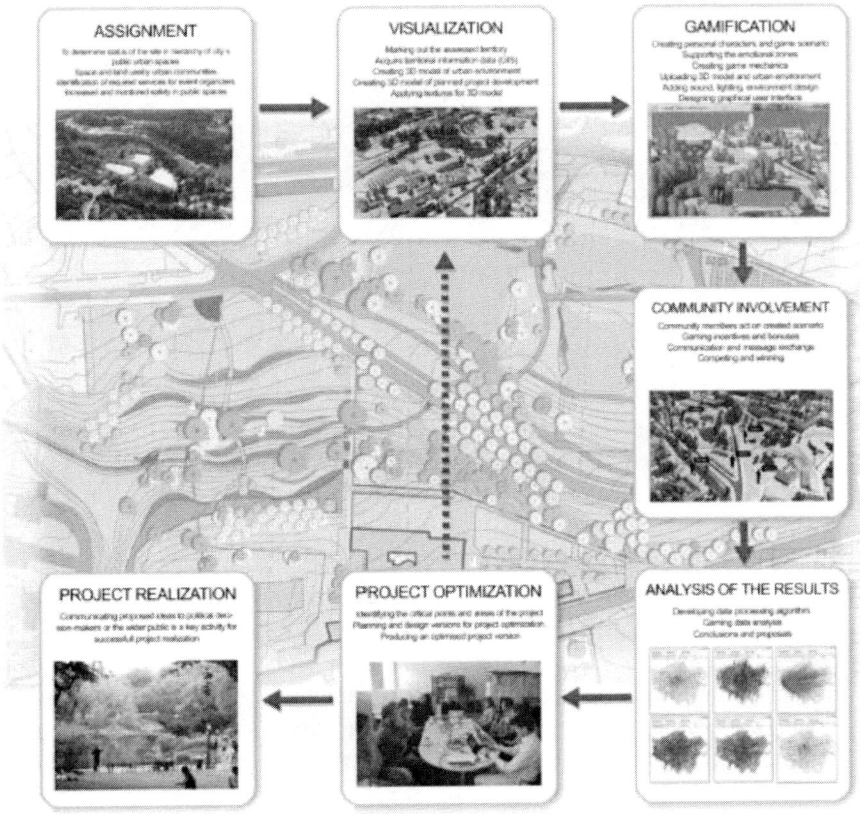

Figure 7: Methodology of virtual urban simulator for public participation. Picture by G. Stauskis.

Participation phase is based on direct involvement of pre-selected residents and other stakeholders in the virtual gaming process by performing pre-designed tasks according to the game scenario. During the starting phase of the project, two scenarios were created: one built around the family recreation activities in the park and the other with recreation of group of friends in the park (Figure 8). In both cases, the users were given certain tools for analysis of the place and performance of the given scenario. In both cases, they could easily move around the park area within enclosed perimeter and select a preferred place for their family or friend's meeting by locating preselected recreation equipment and street furniture at preferred spots of the park area. The scope of tools was designed in a way which could provide the park designers with information about the most and the least attractive places of the park, track the movement of players across the park area to see the busiest tracks, and identify the most necessary public furniture for the park.

a b

Figure 8: Frames of 'The Family' (a) and 'The Friends' (b) scenarios on the VUS models. Unity 3D software environment. Pictures by G. Stauskis.

The analysis phase of VUS includes the data processing and analysis, interpretation of acquired results and their delivery to the project development team for their integration into the final design. The project database will be requested to give a response on data indicators directly related to the tasks of the client that were integrated into the gaming scenarios. The results are expected to put light on the importance, attractiveness and functionality of the planned public space. Finally, the results will be delivered to the planning team for discussions and decisions about the possible actions. In the case of excellent or very good results, the decision might be that the plan could be passed with no or minor additions and amendments. In the case of good or satisfactory results, the plan should be modified by the features pointed out as missing or weak in simulation and testing process. In the case of fair, poor or unsatisfactory evaluation results, the plan has to be seriously reconsidered starting with its aim and objectives, also all planning proposals including socio-economic and environmental planning consequences.

The presented project is still ongoing in its community involvement phase (Figure 7), and the analysis of results will follow in the next phases of research.

RESULTS AND DISCUSSION

The periods of development in public participation illustrate close relations between the historical development of urbanism and growth of civil society. Rapid urbanisation processes brought more people to cities where urban communities with specific lifestyle were developed, and the way how their residential areas were planned and designed gradually became part of social capital with informal rules and traditions of communication, space maintenance, safety and security. In order to keep and grow this value, the built environment of urban communities should be treated very sensitively during the redevelopment or refurbishment phases thus avoiding radical transformations and demolitions. Public participation is the link that provides planning organisers with the information from real living communities allowing to verify and eventually improve the design based on feedback information.

Even though there is a wide range of participation in planning tools and methods established in building legislation and used in planning practice, still there is little experience and research on the application of innovative and emerging public participation techniques, especially with the application of ICT possibilities. Researchers focus on general aspects of evaluation of 'serious games', on hazard recognition [18], environmental [19] and infrastructure spatial planning [20] and other related issues. Traditionally, the set of passive and relatively active methods and techniques prevail in different countries (Table 1). The direct interaction between the planning organiser and the participant in such meetings and committees has relatively low representativeness and reliability, as the events are usually dominated by the most active representatives of communities which usually have strong impact on the feedback from other residents. Usually, the opinions of residents are delivered in an indirect way, as participating activists try to express the opinion of other people who are absent at the taken moment. In this way, the process often has doubtful reliability and low representativeness in respect of users of the planning area. Moreover, the opinions of some groups of population as children, youth or people with disabilities have a rare chance to be heard and minded. In this context, the emerging interactive participation tools as listed in Table 4 are expected to close the mentioned gaps and make the public participation process more complete and its results more reliable and transparent for sustainable urbanism.

Table 4: Methods of reality representation for stakeholder involvement and public participation

Method	Traditional	Modern	Emerging
Graphical	Drawing, drafting, painting	CAD linear drawing	Virtual reality simulation
Spatial	Building hard models, photography	Three-dimensional modelling, model rendering	Gaming, virtual role play, virtual cities

By analysing the methods of public participation, one can notice that each of them has strong sides and weak points; on the other hand, each of them activates a specific area in the community and checks the plan at a special angle. This takes us to an important finding that different tools of public participation should be used in complex by combining methods that are appropriate to the taken case and could give the best available results. To avoid duplication, the most fit-for-purpose methods have to be selected minding cultural, ethical, technical, social, and the other features both in the community and in the project.

In the situation when ICTs are more and more becoming a daily use feature, it is natural to involve them as one of the important tools to be used in public participation. Analysis of recent participation practices demonstrate that ICT-based tools can attract the groups of population that often fall out from the group of participants and give them a modern easily accessible tool using technologies and devices that are already at their disposal. At the very beginning, it should be clear about the strong features and the limitations of this tool as VUS or similar. Virtual reality is not a substitute to the real face-to-face interaction of citizens; it is a supportive tool with several very strong and efficient features as listed in this paper (efficiency, transparency, accessibility and resource consumption) that was missing among traditional instruments of public participation. ICT tools allow creating a multi-layer stakeholder platform that is an essential feature of planning job. Allowing the public into this circle makes the group more complete, and by ICT tools - more powerful in the decisions and the results.

It is essential to have proper software and technical support for the development of this kind of ICT-supported systems. The main dilemma in this phase is finding the best compromise between the systems that are most effective for space modelling on the one side and gaming process on the other. In the VUS case, Unity 3D was chosen as an efficient and comfortable gaming environment that opens wide opportunities in model building and optimisation. Usual drafting, modelling and rendering techniques could be used for the initial phases of the process, later converting the files to a programme supporting virtual gaming. It is essential to keep in order the GIS background and environment so as to get direct references to land sites and buildings in real urban locations.

By responding to the goals of transparency, immediacy, direct access, permanence and follow-up, VUS extends the methods of public participation and introduces a modern, efficient and reliable tool. Essentially, VUS is a new word in the development of interactive and interdisciplinary cooperation platform for public participation that is essential for efficient and effective urban planning and design as foreseen by Agenda 21 and the Aarhus convention in 1998 [21].

CONCLUSIONS

The presented VUS methodology is developed as an innovative tool to help Vilnius City Municipality, on small scale, to examine in an interactive way to what extent the project of Missionaries Park meets the interests of local and all the city's community. The latter will be given an opportunity to get acquainted with the project and report its opinions about all features of the project. Moreover, the method allows checking the planned facility on public attractiveness, functionality, safety and security and, if needed, making the necessary improvements. Lastly, the system enables local authority to establish a contact with potential event organisers and other small businesses in order to shape the park as a space fit for different types of public events that will also be an important feature for economic viability of the place. All together, it leads to more rational use of social and material resources as human labour, urban land, natural and construction materials, and energy.

As permanence and monitoring are important features of planning and design, VUS opens perfect possibilities to monitor the process of planning, construction and maintenance as it could be repeatedly used to indicate the preferred modifications while the facility is in use. There are many other possibilities how the VUS tool could deviate once taken into regular practice by local authorities, communities, businesses and moderated by researchers. Numerous lines of research are open to further development and investigation of specific features of melting public participation with virtual imaging and gaming. As the on-going VUS project already revealed, gaming feature is extremely attractive and involving an environment that could facilitate wider involvement of residents into public participation activities compared to traditional techniques. Interdisciplinary teams of researchers and practitioners are needed to creatively apply the knowledge from the fields of planning and architecture, urban sociology and anthropology, environmental science, ICT and related themes in order to increase the quality of plans and make the implementations more sustainable for the citizens and the environment in the process of construction, use and maintenance.

VUS as an interaction platform can drastically change the design, evaluation, teamwork and participation in planning and design and bring more sustainability instead of hierarchy to the process. VUS application has a high technology edge when the urban space is planned and designed. High visual quality of VUS models is essential to bring realistic familiarity with the realistic landscape image to seriously motivate and involve the user. Finally, the broader questions arise: whether new planning and interaction tools give too much control to the public over their decisions about the future of spaces and buildings usually planned and designed by urban planners and architects. The answers lie in the balanced socio-economic understanding of the roles and missions of architects and the public.

Looking forward, public participation could and should become a trendy exercise to be used actively; joining in ICT technologies is the way to its wider application. Popular and widely used tool would extend the number of users and make it more popular among the groups of young population rarely engaged in ordinary hearings, meetings and committees. In the long run, ICT-based public participation would transfer as a specific application to social media and offer plenty of easily accessible information and effective tools under moderate resource.

AUTHOR'S INFORMATION

Gintaras Stauskis is a Doctor of the Humanities in Architecture and a professor at Vilnius Gediminas Technical University (Lithuania). He is conducting academic training and research activities at the Department of Urban Design, Traku str. 1, Vilnius LT-01332, Lithuania. He is involved in the European Union research programmes on Green Building and sustainable urbanism. He is the Head of Lithuania Green Building Council LTGBC and a member of the editorial board of the *Journal of Civil Engineering and Architecture* (USA), Landscape Architecture and Art (Latvia) and Arhitektura un Pilsetplanošana (Latvia). His current research fields include landscape architecture and planning, 'green' urban architecture and mobility, accessibility of environment, recreation and urban health. His research results have been presented in numerous national and international publications and conferences.

COMPETING INTERESTS

The author declares that he has no competing interests.

REFERENCES

1. Hackney R: *The good, the bad and the ugly*. London: Frederick Muller; 1990.
2. Communities in Control: *Real people, real power*. Crown copyright. Electronic resource; 2008. . Accessed 25 Aug 2013 http://www.openeyecommunications.com/wp-content/uploads/CommunitiesinControlFull.pdf . Accessed 25 Aug 2013
3. Sheppard SRJ: *Visual simulation: a user's guide for architects, engineers and planners*. New York: Van Nostrand Reinhold; 1989.
4. Simpson DM: Virtual reality and urban simulation in planning: a literature review and topical bibliography. *J Plan Lit* 2001, 15: 3.
5. Papasarantou C, Bourdakis V: *Represent-ing presence, V!RUS, n. 8*. 2012. . Accessed 19 August 2013 http://www. nomads. usp.br/virus/virus08/?sec=4&item=4&lang=en . Accessed 19 August 2013
6. Sheffield Development Framework: *Statement of community involvement*. Sheffield: Sheffield City Council; 2006.
7. Statement of Community Involvement and Planning Applications London: Office of the Deputy Prime Minister; 2004. http://www. communities.gov.uk/corporate/ . Accessed 30 July 2013
8. Painting the Landscape: *A cross-cultural exploration of public-government decision-making*. International Association of Public Participation; 2009. . Accessed 25 Aug 2013 http://c.ymcdn.com /sites/www.iap2.org/resource/resmgr/imported/Kettering_FINALExecutiveSummaryReport.pdf . Accessed 25 Aug 2013
9. Visuomenės dalyvavimo teritorijų planavimo procese nuostatai: *[Guidelines on public participation in territory planning]. Lietuvos Respublikos Vyriausybės 1996 m. rugsėjo 18 d. nutarimas Nr.1079 (Lietuvos Respublikos Vyriausybės 2004 m. liepos 16 d. nutarimo Nr. 904 redakcija) [Approved by the Decree of the Government of Lithuania No. 1079 from 18.09.1996 (Version Decree of the Government of Lithuania No. 904 from 16.07.2004)]*. 2004. . 15 May

2013 http://www3.lrs.lt/pls/inter3/dokpaieska.showdoc_l?p_id=237982&p_query=&p_tr2= . 15 May 2013

10. The European Commission: *Directive 2003/35/Ec of the European Parliament and of the Council providing for public participation in respect of the drawing up of certain plans and programmes relating to the environment and amending with regard to public participation and access to justice Council Directives 85/337/Eec and 96/61/Ec*. 2003. . Accessed 25 Aug 2013 http://www3.lrs.lt/pls/inter3/ dokpaieska. showdoc_l?p_id=237982&p_query=&p_tr2= . Accessed 25 Aug 2013

11. Vilnius Gediminas Technical University: *Hotel DZŪKIJA in Druskininkai. Part of the Grand Spa Lithuania complex*. 2013. . Accessed 23 Aug 2013 http://www.construction21.eu/case-studies/lt/hotel-dzkija-in-druskininkai-part-of-the-grand-spa-lithuania-complex.html . Accessed 23 Aug 2013

12. Liggett R, Friedman S, Jepson W: Interactive decision making in a virtual urban world: visual simulation and GIS. In *Proceedings of the 15th ESRI user conference*. Palm Springs; 1995. 24 Feb 2003 24 Feb 2003

13. Bartlett WK, Neumann A, Meifort J: *Interactive landscape planning – results of a pilot study in Koenigslutter am Elm, Germany. Proceedings/Tagungsband; Ed./Hg.: Manfred SCHRENK*. CORP 2005 & Geomultimedia05; 2005. . Accessed 25 Aug 2013 http://www.researchgate.net/publication/228863270_Interactive_Landscape_PlanningResults_of_a_pilot_study_in_Koenigslutter_am_Elm_Germany . Accessed 25 Aug 2013

14. Ruffat S, Ter Minassian H: Video games and urban simulation: New tools or new tricks? *Eur J Geogr* 2012. article no. 622. . Accessed 25 Aug 2013 http://cybergeo.revues.org/25561?lang=en article no. 622. . Accessed 25 Aug 2013

15. Dhingra Ch KRSR: *Establishing a multi-stakeholder forum for urban mobility*. EMBARQ, Washington: UN-Habitat; 2013.

16. Manneh C: Computer game provides inspiration for new planning tool. *Cities Today* 2012, 7: 43.

17. Bučys J: Funkcinių Ir kompozicinių urbanistinės struktūros analizės aspektų sujungimas: vilniaus miesto lokalių centrų tyrimas. [Combining functional and compositional aspects of the analysis of urban structure: a case study of local centres in Vilnius]. *J Arch Urbanism* 2013, 37(2):69–87. doi:10.3846/20297955.2013.807565

18. Mayer IS, Wolff A, Wenzler I: Learning efficacy of the 'hazard recognition' serious game: a quasi experimental study. In *Serious*

games development and applications. Heidelberg: Springer; 2013:118–129.

19. Nefs M, Gerretsen P, Dooghe D, Mayer IS, Meijer SA: Gaming the interrelation between rail infrastructure and station area development: part 1 - modeling the serious game 'SprintCity'. *Third international conference on infrastructure systems and services: next generation infrastructure systems for EcoCities* 2010. 11–13 Nov 2010 11–13 Nov 2010

20. Zhou Q, Mayer IS: Gaming as the method to integrate modelling and participatory approaches in interactive water management. In *International Environmental Modelling and Software Society (iEMSs) 2010 International congress on environmental modelling and software modelling for environment's sake, fifth biennial meeting*. Ottawa: International Environment Modelling and Software Society iEMSS; 2010.

21. United Nations Economic Committee for Europe: *Convention on access to information, public participation in decision-making and access to justice in environmental matters*. 1998. . Accessed 25 Aug 2013 http://www.unece.org/fileadmin/DAM/env/pp/documents/cep43e.pdf . Accessed 25 Aug 2013

13

GEOGRAPHICAL ANALYSIS OF COMMUNITY RESILIENCE TO SEISMIC HAZARD IN SOUTHWEST CHINA

Xiaolu Li[1], Lei Wang[2], Shan Liu[1]

[1] School of Automation, University of Electronic Science and Technology of China, Chengdu 611731, Sichuan, China

[2] Department of Geography and Anthropology, Louisiana State University, Baton Rouge, LA 70810, USA

ABSTRACT

This article presents an explorative analysis of community resilience to seismic hazard in the 2008 Wenchuan Earthquake area of Southwest China. We used a regression model to analyze the impact of 13 key socioeconomic and demographic variables on community resilience in 105 counties, based on data derived from population census and provincial statistical yearbooks of China. In this research, we argue that community resilience should be measured by the change of population growth rate (Δdp) instead of population growth rate (dp) when using socioeconomic data from a fast-growing country such as China. Using Δdp as the dependent variable resulted in a better regression model. To avoid the common multicollinearity problems among the independent variables, a principal component-based factor analysis was used to consolidate the socioeconomic variables into four comprehensive factors. The geographically weighted regression coefficient maps revealed the spatial pattern of the association of the variables with resilience. We also used the K-means cluster method to segment the study area into four subregions that exhibit localized characteristics defined by the regression coefficients. In this way, we could infer location-sensitive disaster management policies that help to enhance social resilience to seismic hazards.

Keywords: China, Community resilience, Geographic information system, Geographically weighted regression, Wenchuan Earthquake

1 INTRODUCTION

In a rapidly developing country like China, society becomes more vulnerable to seismic hazards because of the relatively low preparedness for disasters in comparison with the high speed growth of economy and population accompanied by fast urban expansion (Zheng et al. 2015). The 2008 Wenchuan Earthquake in Southwest China has attracted significant attention because it caused great casualties and substantial property losses (Li et al. 2015a). Faced with potential future tragic losses from seismic disasters, China needs to address scientific questions such as: (1) Will severe losses slow down the recovery of the affected communities? (2) If not, what are the social and economic characteristics that would contribute most to post-disaster recovery and enhance the impacted community's resilience? (3) How should the government develop locally adapted strategic plans for optimized regional recovery with limited resources?

The definition of social resilience has been discussed in the literature (Holling 1973, 1996; Adger et al. 2005; Cutter et al. 2008). Turner et al. (2003) have pointed out that vulnerability is not only just exposure to hazards, but also includes sensitivity and resilience to disastrous events. Some researchers (such as Norris et al. 2008) have suggested that resilience can be understood as a "capacity," a "process," an "outcome," or a certain combination of these three concepts for reducing damage and improving recovery. Brown and Kulig (1996) used community resilience to measure different communities' recoverability or adjustability from misfortune. Paton and Johnston (2001) defined community resilience as the capacity to recover from hazard damage and the ability to effectively deal with future exposure to hazards. We adopt the definition of resilience proposed by the United States National Research Council (NRC 2012, p. 1), which considers resilience to be "...the ability to prepare and plan for, absorb, recover from, and more successfully adapt to adverse events over time." Through an explorative analysis using regression models, this research identifies the economic and social characteristics of disaster-stricken communities that could affect their post-disaster recovery. We anticipate that the scientific findings from this research can help government agencies and policymakers formulate quick response strategies to seismic disasters so a disaster area can quickly return to its normal status, or in other words, achieve greater resiliency.

Previous research has suggested that a variety of statistical models can successfully evaluate the association of resilience and socioeconomic variables. For example, Cutter et al. (2008) provided a theoretical framework for local scale resilience assessment—the disaster resilience of place model. These authors treated resilience as a continuous dynamic process influenced by both exposure and exogenous factors from social systems. Later Cutter et al. (2010) introduced another set of indicators to derive a baseline resilience index for communities. Lam et al. (2015) used the resilience inference model to assess community resilience to hurricane hazards. Bruneau et al. (2003) developed a quantitative framework that measured the speed of recovery using four resilience dimensions (technical, organizational, social, and economic) of four systems (electric power, water, hospital, and local response and recovery systems). Based on this framework, they evaluated how economic losses would affect resilience.

Many social science researchers have criticized quantitative methods because numeric measures often neglect the uniqueness of a society or its individual members. Such uniqueness requires that the policies inferred from simulation models must be location sensitive (Cloke et al. 1991; Smith 1998). In most applications of geographical analyses, statistical approaches are used in a "cookbook" fashion without considering the spatial features of the data (Fotheringham et al. 2000). General nonspatial statistical models do not incorporate locational information in their calculations, and therefore they are not suitable for policy making under conditions of geographical heterogeneity (Zheng et al. 2016). Although the important role of geographic information systems (GIS) in disaster management sciences is widely accepted, most research has only used GIS as a mapping and data collection tool (Zheng et al. 2013), and has neglected the ability of GIS to provide spatially varied disaster mitigation plans and recovery strategies. In this research, we have introduced a geographically weighted regression (GWR) model supported by a GIS in order to build location-dependent regression models. The aim is to find a better strategic approach to community recovery and sustainable socioeconomic development in earthquake disaster areas. GWR is a local regression model that could reconcile the use of statistical regression models with GIS (Fotheringham et al. 1998, 2003). The use of locally sampled data and distance-weighted solutions addresses the spatial nonstationarity problem of regression models. In addition, the regression coefficients from GWR exhibit a certain spatial pattern, the correct reading of which could help better understand the association of resilience with socioeconomic variables place by place (Brunsdon et al. 1998). We can derive location-specific disaster management plans that effectively assist to enhance the sustainability of those areas that are threatened by seismic hazards.

2 STUDY AREA AND DATA SOURCES

The study area is the region in southwestern China that was impacted by the Wenchuan Earthquake in 2008, including the epicenter and surrounding areas in Sichuan Province (Fig. 1a). The Wenchuan Earthquake caused over 69,227 casualties and resulted in losses of RMB 845.1 billion Yuan (over USD 130 billion); more than 90 % of these losses occurred in Sichuan Province (Guo 2012). The quake intensity map in Fig. 1b was based on the data provided by the United States Geological Survey (USGS 2008). As shown in Fig. 1b, the earthquake intensity peaked at the epicenter in Wenchuan and Lixian Counties, and gradually weaker impacts occurred up to about 200 km away. Figure 1c shows the area with major damage and losses caused by the earthquake, measured in loss per capita derived from the 2008 Sichuan Statistical Yearbook (Sichuan Provincial Bureau of Statistics 2008).

Demographic and socioeconomic data used in the study were selected from the 2000 population census of the People's Republic of China (National Bureau of Statistics of the People's Republic of China 2001) and the Sichuan Statistical Yearbooks of 2003 and 2008 (Sichuan Provincial Bureau of Statistics 2003, 2008). Many indicators for the measurement of community resilience are found in the disaster risk literature. Paton et al. (2000, 2006) found that the demographic characteristics of the community, including age, race, gender, occupation, and socioeconomic condition, were related to social resilience. Cutter et al. (2008) assessed community resilience according to economic indicators and community competence, such as employment, value of property, wealth generation, health and wellness, and so on. The H. John Heinz III Center for Science, Economics, and the Environment (2000) concluded that social vulnerability and resilience were influenced by several socioeconomic characteristics including commercial and industrial development, residential property, level of education, and medical services, among others. But there has not been a consensus on the best core variables with which to study community resilience.

(a)

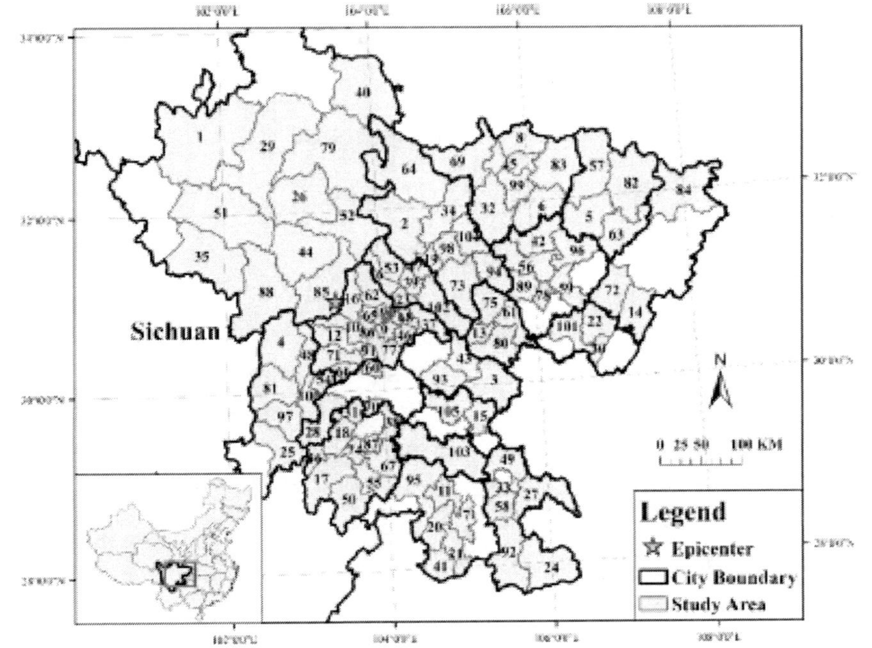

1	Aba	20	Gaoxian	42	Langzhong	64	Pingwu	85	Wenchuan
2	Anxian & Beichuan	21	Gongxian	43	Lezhi	65	Pixian	86	Wenjiang
		22	Guanganqu	44	Lixian	66	Pujiang	87	Wutongqiao
3	Anyue	23	Guanghan	45	Lizhou	67	Qianwei	88	Xiaojin
4	Baoxing	24	Gulin	46	Longquanyi	68	Qingbaijiang	89	Xichong
5	Bazhou	25	Hanyuan	47	Luojiang	69	Qingchuan	90	Xindu
6	Cangxi	26	Heishui	48	Lushan	70	Qingshen	91	Xinjin
7	Changning	27	Hejiang	49	Luxian	71	Qionglai	92	Xuyong
8	Chaotian	28	Hongya	50	Mabian	72	Quxian	93	Yanjiang
9	Chengdu Metropolitan Area	29	Hongyuan	51	Maerkang	73	Santai	94	Yanting
		30	Huaying	52	Maoxian	74	Shawan	95	Yibin
		31	Jiajiang	53	Mianzhu	75	Shehong	96	Yilong
10	Chongzhou	32	Jiange	54	Mingshan	76	Shifang	97	Yingjing
11	Cuiping	33	Jiangyang	55	Muchuan	77	Shuangliu	98	Youxian
12	Dayi	34	Jiangyou	56	Nanbu	78	Shunqing	99	Yuanba
13	Daying	35	Jinchuan	57	Nanjiang	79	Songpan	100	Yucheng
14	Dazhu	36	Jinkouhe	58	Naxi	80	Suining Urban Area	101	Yuechi
15	Dongxing	37	Jintang	59	Pengan			102	Zhongjiang
16	Dujiangyan	38	Jinyan	60	Pengshan	81	Tianquan	103	Zigong
17	Ebian	39	Jinyang	61	Pengxi	82	Tongjiang	104	Zitong
18	Emeishan	40	Jiuzhaigou	62	Pengzhou	83	Wangcang	105	Zizhong
19	Fucheng	41	Junlian	63	Pingchang	84	Wanyuan		

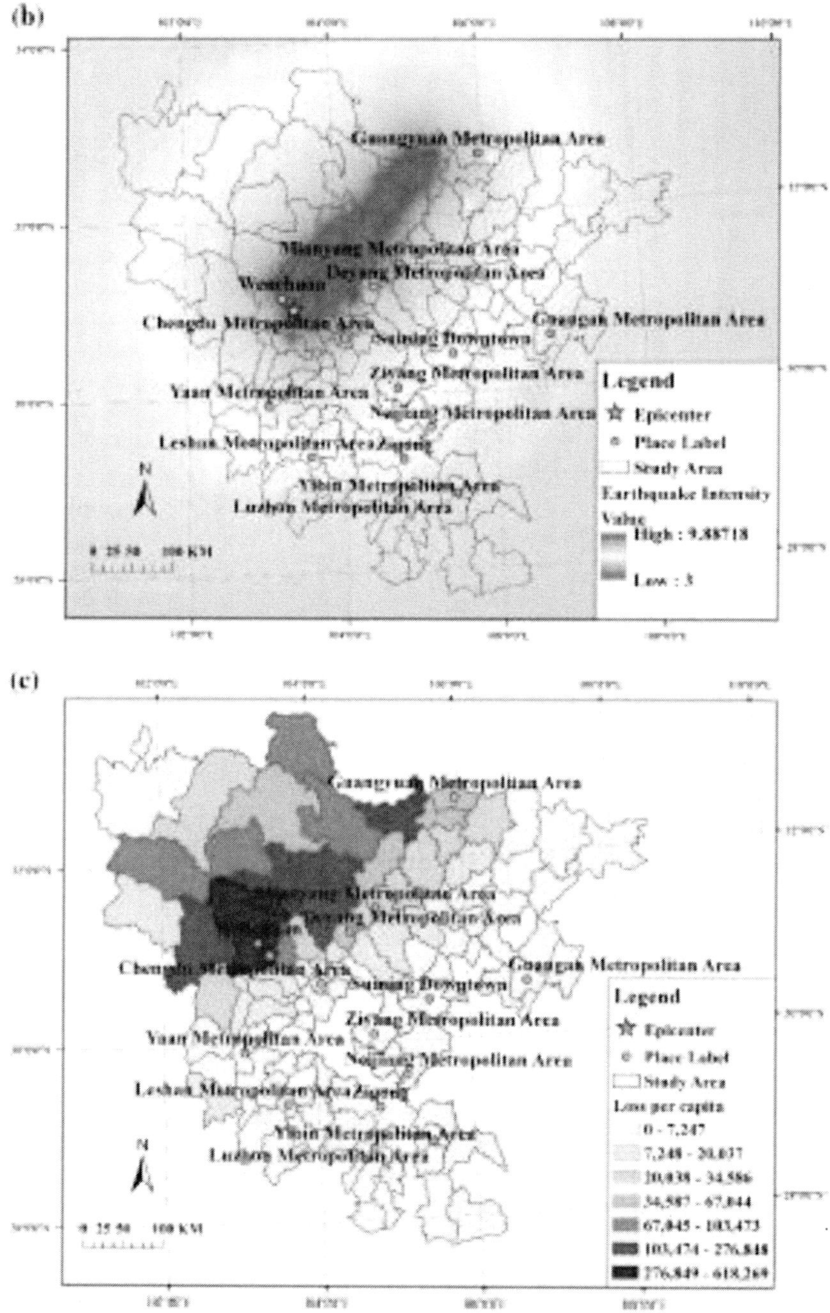
(b)
Guangyuan Metropolitan Area
Mianyang Metropolitan Area
Deyang Metropolitan Area
Wenchuan
Chengdu Metropolitan Area
Guangan Metropolitan Area
Suining Downtown
Ziyang Metropolitan Area
Yaan Metropolitan Area
Neijiang Metropolitan Area
Leshan Metropolitan Area
Zigong
Yibin Metropolitan Area
Luzhou Metropolitan Area
N
Legend
Epicenter
Place Label
Study Area
Earthquake Intensity Value
High : 9.88718
Low : 3
(c)
Guangyuan Metropolitan Area
Mianyang Metropolitan Area
Deyang Metropolitan Area
Chengdu Metropolitan Area
Guangan Metropolitan Area
Suining Downtown
Ziyang Metropolitan Area
Yaan Metropolitan Area
Neijiang Metropolitan Area
Leshan Metropolitan Area
Zigong
Yibin Metropolitan Area
Luzhou Metropolitan Area
N
Legend
Epicenter
Place Label
Study Area
Loss per capita
0 - 7,247
7,248 - 20,037
20,038 - 34,586
34,587 - 67,044
67,045 - 103,473
103,474 - 276,848
276,849 - 618,269

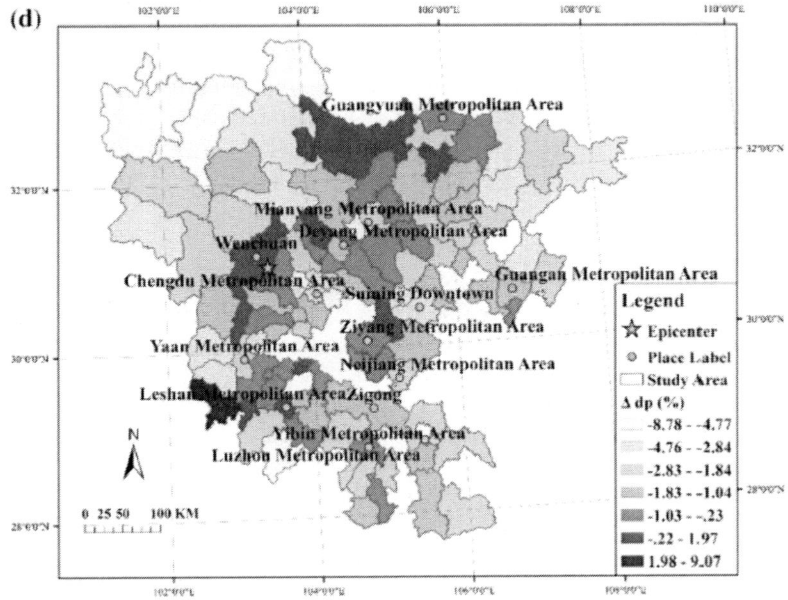

Fig. 1: **a** The study area and the selected counties with data available, **b** intensity of the 2008 Wenchuan Earthquake interpolated from USGS earthquake dataset, **c** economic losses per capita of the counties, and **d** change of population growth rate of the counties (Δ*dp*)

In light of previous vulnerability and resilience research (The H. John Heinz III Center for Science, Economics, and the Environment 2000; Paton et al. 2000; Cutter et al. 2003; Paton and Johnston 2006; Nelson et al. 2009), and considering the limited data availability in China, we selected 13 social and economic variables for our regression analysis, as listed in Table 1. Social condition was accounted for using total population, ethnic and urban populations, age, education levels, and employment variables. To account for the level of economic development, GDP per capita, proportion of primary industry, and proportion of secondary industry were selected to reflect economic influences on resilience. Additionally, we used average wage, per capita saving deposit balances, and the number of hospital and social welfare beds as the index of the level of living standard, medical treatment, and social welfare for each county. We opted to normalize the data by total population to avoid the modifiable areal unit problem (Fotheringham et al. 1995). Besides the variables in Table 1, the total population data of 2005, 2007, 2008, and 2010 for each county were collected from the annual Sichuan Statistical Yearbooks (Sichuan Provincial Bureau of Statistics 2006, 2008, 2009, 2011) to calculate the change of population growth rate (Δ*dp*) before and after the earthquake (Fig. 1d). Not all counties within the earthquake area were included in the analysis because of the missing data problem. For this reason, we merged five counties (Jinjiang, Jinniu, Qingyang, Wuhou, and Chenghua) into one unit of analysis as the Chengdu Metropolitan Area. We used the aggregated data of the Suining urban areas, instead of its two subareas: Chuanshan and Anju Counties. Furthermore, six counties of Zigong City experienced administrative boundary change between 2000 and 2010. Hence we used Zigong City as one study unit. For the same reason, we treated Beichuan and Anxian Counties as one unit. To ensure that the study area is contiguous, a few geographically isolated counties were not included, such as Kaijiang and Tongchuan Counties. As a result, a total of 105 units (Fig. 1a) from Sichuan Province were used for data analyses.

Table 1 Variables used in the study

No.	**Labels**	**Socioeconomic variables**	**Variable meaning**	**Units**	**Sources**
Demographic					
1	PopDensity	Population density, 2007	Population	Persons/km^2	YB
2	RtoEthPop	Percentage of ethnic minorities population, 2000	Ethnicity	%	CS
3	RtoUrbanPop	Percentage of urban population, 2007	Urban	%	YB
4	RtoPopAge 15–64	Percentage of population aged 15–64, 2000	Age	%	CS
Social					
5	RtoEduSec	Percentage of population	Education	%	CS

No.	Labels	Socioeconomic variables	Variable meaning	Units	Sources
		with education of senior secondary school and technical secondary school and above, 2000			
6	RtoEmpPop	Ratio of employed population, 2007	Employment	%	YB
Economic					
7	GDPperCapita	GDP per capita, 2007 (at current price)	Commercial and industrial development	Yuan/person	YB
8	PPriIndus	Proportion of primary industry, 2007	Commercial and industrial development	%	YB
9	PSecIndus	Proportion of secondary industry, 2007	Commercial and industrial development	%	YB
10	AvWages	Average wages of staff and workers, 2007	Residential property	Yuan/person	YB

No.	Labels	Socioeconomic variables	Variable meaning	Units	Sources
11	PCSvgsDpstB	Per capita savings deposit balances of residents, 2002	Residential property	Yuan/person	YB
Health					
12	PCHospBed	Number of hospital beds per 10,000 persons, 2007	Medical capacity	Unit/10,000 person	YB
Social welfare					
13	PCSWBed	Number of social welfare home beds per 10,000 persons, 2007	Social welfare	Unit/10,000 person	YB

The 2000 population census data were obtained from the National Bureau of Statistics of the People's Republic of China (CS) and the provincial statistical yearbooks 2003 and 2008 (YB)

3 DATA ANALYSES

In this section, we present the data analysis of the socioeconomic variables. First, we calculated the change in population growth rate (Δdp) as the dependent variable. Then, we used the factor analysis to consolidate the explanatory variables into four factors. A global ordinary least squares (OLS) regression model was used to confirm the association between the dependent variables and the explanatory variables. The following GWR revealed the spatial dependence of the regression coefficients. We also carried out a K-means clustering analysis on the GWR coefficients to derive location-sensitive policies that would, if employed, enhance community resilience.

3.1 Resilience Index

In the hazard risk research literature, no consensus exists on what is the best way is to measure resilience (Bevington et al. 2011). Some scholars advocate the use of evacuation/return ratio as a resilience measurement. However, the data for tracking population migration are not often available, especially for large study area like the one examined in this research. Even in a data-rich area such as New Orleans in the US after Hurricane Katrina in 2005, it is still impossible to track people's migration. Therefore, we chose to use the change in population growth rate (Δdp) before and after the earthquake as the resilience index in this study. Population is a more summative variable to measure resilience than economic variables, such as GDP and income growth (Pelling 2003; UNISDR 2004; Cutter et al. 2008). Population growth (*dp*) has been similarly used by many studies as the recovery indicator (Chang 2010; Finch et al. 2010; Li et al. 2010, 2015a). Nevertheless, this measure of population growth may include the portion from natural growth, which might have a higher contribution than the impact of disasters, especially in areas experiencing rapid urban expansion. For this reason, we used the measure of change in population growth rate between the pre- and post-seismic event periods as the resilience index, which is calculated as:

$$\Delta dp = \frac{P2010 - P2008}{P2008} - \frac{P2007 - P2005}{P2005}, \tag{1}$$

where Δdp is the difference of population growth rate before and after the 2008 Wenchuan Earthquake event, P2010 is the population of 2010, and so on. Figure 1d shows the spatial pattern of the change of population growth rate (Δdp). We assume that the population growth rate was mainly affected by the earthquake disaster, that is, deaths, migration, and immigration. We make this assumption based on the following reasoning: (1) population growth rate is relatively stable in China unless there is an abrupt change in policies, such as birth control or fertility, (2) if the first assumption holds true, the change in population growth rate of the area during the 2005–2010 period should be associated with the major disaster event of 2008, and (3) communities with high resilience should maintain a relatively stable population growth rate when hit by such disasters.

3.2 Factor Analysis

For independent variables we use 13 socioeconomic variables. The descriptive statistics of these socioeconomic variables are shown in Table 2. Because the socioeconomic variables are always correlated, they must be selected or transformed to avoid the multicollinearity problem (Ghosh and Manson 2008). In this research, we employed principal component-based factor analysis to create a set of new variables that are orthogonal to each other. Factor analysis can reduce the dimensionality of the data by aggregating them to fewer comprehensive factors, reduce the possibility of multicollinearity existing in the variables, and at the same time maintain sufficient data variance with more comprehensive factors (Harman 1976; Chatfield and Collins 1980; Hair et al. 1998). Therefore, factor analysis is quite often used on multivariate data for dimension reduction or to condense multiple input variables to several orthogonal and meaningful factors (Hair et al. 1998). It is widely used for dealing with a large number of input variables that are potentially correlated to each other at a certain level. For example, many social studies have used factor analysis to derive a social vulnerability index (Cutter et al. 2003) for disaster science and management. Unlike the piecewise regression method that discards some input variables, factor analysis tends to retain useful information as much as possible from all variables. It also helps remove data noise by extracting a few principal components as factors, neglecting those components with low eigenvalues, which are possible noise components.

Table 2 Descriptive statistics of socioeconomic variables for the study

	Minimum	**Maximum**	**Mean**	**Standard deviation**
PopDensity	4.88	6542.92	510.83	676.38
RtoEthPop	0.02	94.11	9.58	23.72
PtoPopAge15–64	61.75	80.28	70.11	3.51
PtoEduSec	3.89	40.98	9.66	5.51
RtoUrbanPop	5.77	100.00	23.74	15.02

	Minimum	Maximum	Mean	Standard deviation
RtoEmpPop	26.24	129.68	57.43	11.13
PPriIndus	0.17	52.12	25.78	12.37
PSecIndus	13.38	82.86	43.29	14.51
GDPperCapita	4645.00	50,246.49	12,927.76	7997.90
AvWages	12,997.00	31,207.00	18,647.42	4448.45
PCSvgsDpstB	765.05	21,981.52	3352.90	2723.11
PCHospBed	0	71.23	24.27	14.16
PCSWBed	0	52.24	13.84	8.84

We conducted the factor analysis in the form of a principal component analysis (PCA) (Jolliffe 2010). Thirteen socioeconomic variables were put into the factor analysis module in the SPSS statistical software package. The variables were standardized to zero mean and unit variance before being transformed. Feasibility tests using the Kaiser–Meyer–Olkin (KMO) test for sampling adequacy and Bartlett's sphericity test indicate the suitability of the variables for factor analysis, which can be found in Table 3. The KMO value of greater than 0.8 shows that the correlation among the variables is relatively high. The Bartlett's test indicates that the variables' independence assumption should be rejected, and the application of factor analysis was necessary. As a result, the 13 input variables were reduced to four factors by PCA and a varimax rotation (Table 4), which account for 79.95 % of the variance in the input variables. The variables with high component-loading scores (red-colored numbers in Table 4) were used to interpret the factors. The factors were identified as follows: Factor 1 is about urbanization level (population density, age, level of education, and urban population, residential property, and medical care capacity), Factor 2 is defined as economic structure, Factor 3 is income level and ethnicity, and Factor 4 is about employment status and social welfare.

Table 3: The result of KMO and Bartlett's test

Kaiser–Meyer–Olkin measure of sampling adequacy	0.83
Bartlett's test of sphericity	
Approx. Chi square	1219.91
Df	78.00
Sig.	0

Table 4: Rotated component matrix result

	Component			
	1	2	3	4
PopDensity	0.85	-0.20	-0.20	0.14
RtoEthPop	-0.21	-0.03	0.90	-0.07
PtoPopAge15-64	0.69	0.41	-0.13	0.09
PtoEduSec	0.89	0.30	0.16	-0.03
RtoUrbanPop	0.87	0.32	0.12	0.08
RtoEmpPop	-0.01	0.40	0.03	0.59
PPriIndus	-0.53	-0.73	-0.19	-0.09
PSecIndus	0.15	0.88	-0.11	0.08
GDPperCapita	0.75	0.51	0.09	0.24
AvWages	0.48	0.06	0.68	0.23
PCSvgsDpstB	0.94	0.20	0.05	0.06
PCHospBed	0.53	0.51	0.37	0.09
PCSWBed	0.17	-0.05	0.02	0.88

Extraction method: principal component analysis, rotation method: varimax with Kaiser normalization

Red-colored numbers represent the highest component-loading scores for each variable

The maps of the four factors for the 105 counties are shown in Fig. 2. From Fig. 2a, we can see that the counties with high Factor 1 score are mainly in highly urbanized metropolitan areas, such as the metropolitan areas of Chengdu, Mianyang, Deyang, Yibin, Yaan, and Luzhou, and the counties surrounding these urban areas (Fig. 2a). The map of Factor 2 (Fig. 2b) shows economic dependency on secondary industries and those areas with secondary industries as the mainstay of their economic base (Wenchuan, Lixian, Maoxian, Yaan, and so on) are mainly found in the west of the region. The counties with primary industries are mostly located in the east. The Chengdu Metropolitan Area has the lowest scores of Factor 2 in the study area—in Chengdu, the proportion of the secondary industries in GDP is very low, and the mainstay of the economic base is the tertiary industry. Factor 3 (Fig. 2c), which combines income and ethnicity, exhibits the distribution of the ethnic minority people and higher income in the 105 counties. The ethnic minority population resides predominantly in the Aba Prefecture. This area is relatively sluggish in economic development due to its high topographic relief and physical isolation from other areas. Two other prefectures in Sichuan Province—Ganzi and Liangshan—are also mainly occupied by minority population. But due to missing county-level economic loss data, Ganzi and Liangshan are not included in our study, which makes Aba Prefecture a geographically isolated area with respect to Factor 3. In Fig. 2d, areas with a high Factor 4 score (employment and social welfare) are located around the metropolitan areas of Chengdu, Deyang, Ziyang, Leshan, and Guangan.

3.3 Confirmative Analysis with Global Ordinary Least Squares Model

The four factors from the factor analysis along with the two variables about the earthquake—earthquake intensity and loss per capita—were used as the explanatory variables in a linear regression to examine their relationships with the change in the population growth rate. The change in population growth rate before and after the earthquake event (Δdp) was used as the resilience index, that is, the dependent variable of the regression.

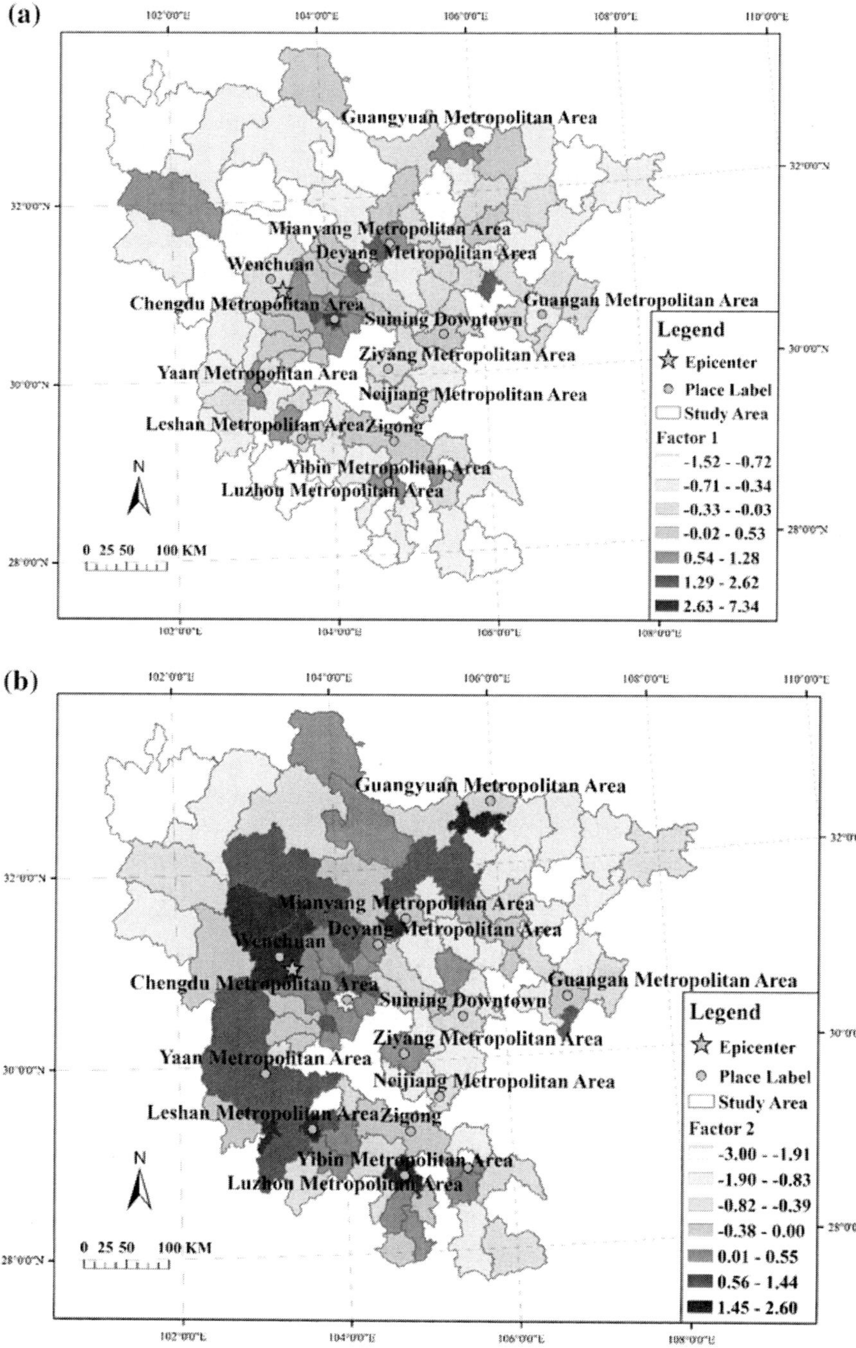
(a)
Guangyuan Metropolitan Area
Mianyang Metropolitan Area
Deyang Metropolitan Area
Wenchuan
Chengdu Metropolitan Area
Suining Downtown
Guangan Metropolitan Area
Ziyang Metropolitan Area
Yaan Metropolitan Area
Neijiang Metropolitan Area
Leshan Metropolitan Area
Zigong
Yibin Metropolitan Area
Luzhou Metropolitan Area
N
0 25 50 100 KM
Legend
Epicenter
Place Label
Study Area
Factor 1
-1.52 - -0.72
-0.71 - -0.34
-0.33 - -0.03
-0.02 - 0.53
0.54 - 1.28
1.29 - 2.62
2.63 - 7.34
(b)
Guangyuan Metropolitan Area
Mianyang Metropolitan Area
Deyang Metropolitan Area
Wenchuan
Chengdu Metropolitan Area
Suining Downtown
Guangan Metropolitan Area
Ziyang Metropolitan Area
Yaan Metropolitan Area
Neijiang Metropolitan Area
Leshan Metropolitan Area
Zigong
Yibin Metropolitan Area
Luzhou Metropolitan Area
N
0 25 50 100 KM
Legend
Epicenter
Place Label
Study Area
Factor 2
-3.00 - -1.91
-1.90 - -0.83
-0.82 - -0.39
-0.38 - 0.00
0.01 - 0.55
0.56 - 1.44
1.45 - 2.60

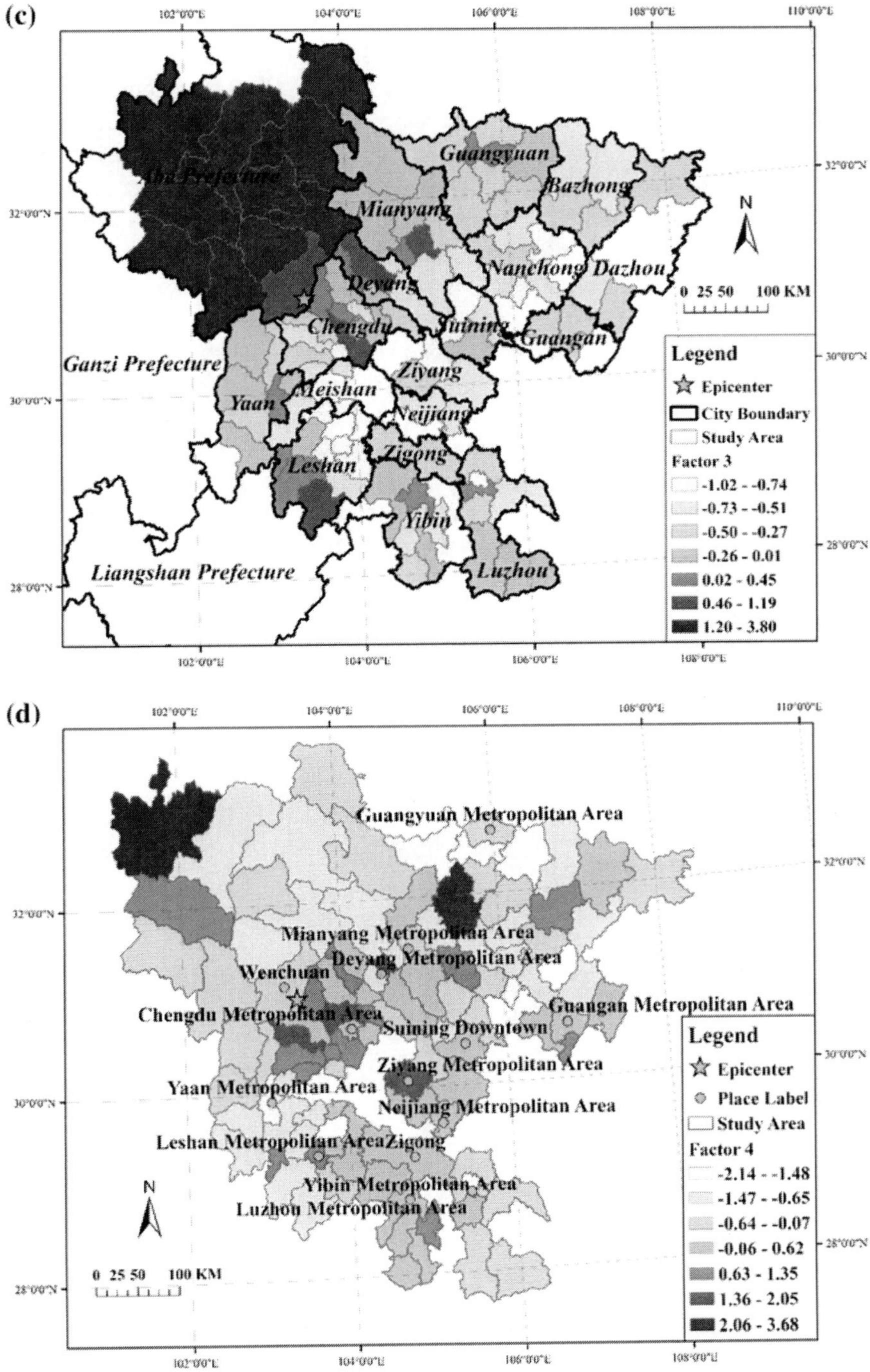

Fig. 2: Extracted factors from PCA and varimax rotation: **a** Factor 1, **b** Factor 2, **c** Factor 3, and **d** Factor 4

The initial round of OLS regression showed there are two outliers: Hanyuan County (OLS residual 10.66) and Mabian County (OLS residual −6.30). Because regression models can be severely affected by outliers, we removed these two counties and used only the remaining 103 units in the study area in the following analysis. At the second round of regression with the 103 counties, the least squares regression result (Table 5) shows that the intercept (general level of recovery capability), loss per capita, and Factors 1 and 3 in the model are statistically significant. Loss per capita has a positive coefficient, whereas the intercept and Factors 1 and 3 have negative coefficients. The variance inflation factor (VIF) values, which are used to quantify the severity of multicollinearity in the OLS analysis, do not suggest any multicollinearity among the independent variables. The coefficient of adjusted r-squared for the recovery capability is 0.31, which suggests a significant amount of unexplained variance by the regression model. The residual map (Fig. 3) shows that the residual has no obvious spatial clusters. The negative sign of the intercept indicates that in general this study area had a significantly lower population growth rate after the earthquake than the pre-seismic event rate. This change was very likely due to the earthquake disaster. The negative sign of Factor 1 indicates that the resilience to the earthquake disaster may be inversely related to the degree of urbanization. Areas with a higher density of human construction may need more time to rebuild than the rural areas.

Table 5: Ordinary least squares regression statistical report

Variables	**Unstandardized coefficients**		**Standardized coefficients**	**Sig.**	**Robust_Pr**	**VIF**
	B	**Standard error**	**Beta**			
Intercept	−2.5082	1.0427		0.02*	0*	
Intensity	0.1026	0.1905	0.0585	0.59	0.51	1.76
Loss per capita	4.25E−06	1.93E−06	0.2618	0.03*	0.01*	2.07
Factor 1	−0.3633	0.1322	−0.2446	0.01*	0*	1.17

Variables	Unstandardized coefficients		Standardized coefficients	Sig.	Robust_Pr	VIF
	B	Standard error	Beta			
Factor 2	0.2550	0.1341	0.1688	0.06	0.11	1.14
Factor 3	−0.7472	0.1379	−0.5039	0*	0*	1.26
Factor 4	0.0368	0.1272	0.0246	0.77	0.78	1.06
Akaike's information criterion	344.87					
Adjusted r-squared	0.31					

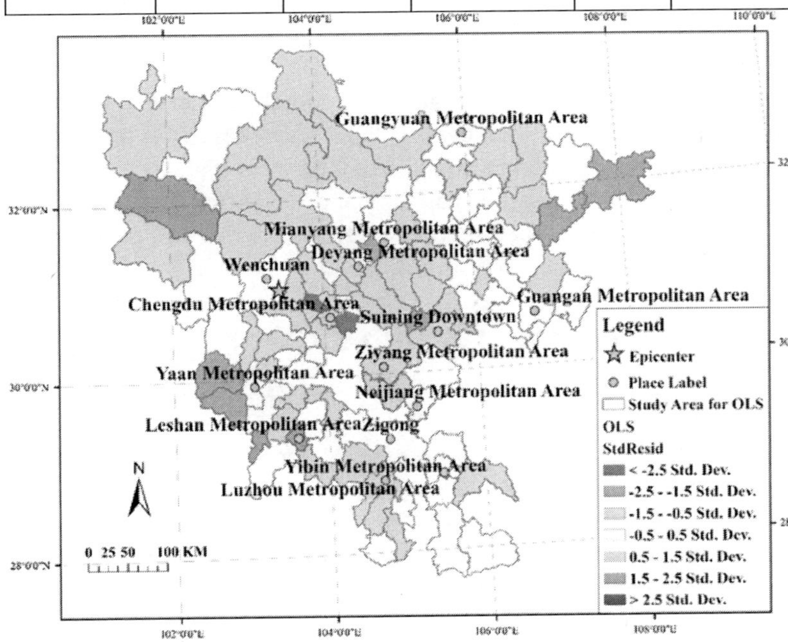

Fig. 3: Residuals of the ordinary least squares regression model

The coefficient of Factor 3 is also significantly negative, suggesting that the ethnic minorities have less resilience to earthquakes. Similar findings about ethnic minority groups were reported before but in different study areas in metropolitan areas in the eastern and southern US (Cutter et al. 2003). In fact, many scholars have suggested that cultural obstacles and marginalization of ethnic minorities would have an adverse effect on these communities' social vulnerability, resistance to disasters, and the ability to recover (Cutter et al. 2003). In Southwest China, a high percentage of ethnic minority communities often live in isolated areas, are poor with limited income and low levels of economic development, and exhibit a lack of infrastructure as well as limited access to communication technology (Shan 2010).

According to the disaster literature (Cutter et al. 2003; Bruneau et al. 2003), areas with more damage and greater economic loss may need a longer time to recover. This is not the case with the Wenchuan Earthquake area in Southwest China, however, as illustrated by our regression model. Loss per capita from the earthquake has a significant positive association with resilience in our case study. This apparent counterintuitive result can be explained by the fact that those counties that suffered more losses have received more resources from the government for reconstruction. The prioritized allocation of resources for reconstruction, labor, and funds from society and government might have contributed to a faster recovery and thus created apparent higher community resilience from the seismic disaster. Economic loss per capita is still the main criteria for prioritizing disaster relief and recovery by the government (Guo 2012). Our model shows that communities and individuals who really need more assistance may well be located in ethnic monitory and densely built-up areas. The standardized coefficients of the independent variables ranked from high to low by their magnitude are: Factor 3, loss per capita, Factor 1, Factor 2, intensity, and Factor 4. Because Factor 2, Factor 4, and intensity are not significant, we do not provide deeper discussion on these variables.

3.4 Explorative Analysis with the Geographically Weighted Regression Model

In addition to the classic OLS regression model, localized regression models such as a GWR could provide knowledge about the spatial stationarity of the model and spatial pattern of the coefficients (Unwin 1996; Fotheringham et al. 1998). Unlike the OLS method, which estimates the global behavior of coefficients regardless of geographical settings and locations (Panik 2009), GWR estimates the local parameters β^β^(*i*) at location *i* that minimize local regression error constrained by a geographically weighted matrix w_i (Hastie and Tibshirani 1990; Fotheringham et al. 1998, 2003; Panik 2009):

$$\hat{\beta}(i) = \left(X^T W(i) X\right)^{-1} X^T W(i) y \quad , (2)$$

where $W(i) = \begin{pmatrix} W_{i1} & 0 & \cdots & 0 \\ 0 & W_{i2} & \cdots & 0 \\ \vdots & \vdots & \ddots & \vdots \\ 0 & 0 & \cdots & W_{in} \end{pmatrix}$ is a matrix of weights specific to location *i* and the values are between 0 and 1. W_{in} is the weight given to data point *n* by the impact between location *i* and location *n*. The weights are usually modeled as a Gaussian-shape distance decay function with higher weights assigned to closer locations (Fotheringham et al. 2003).

We ran the GWR analysis with the same variables selected for the global OLS model. The bandwidth parameter of the GWR model is set to be adaptive. The adaptive bandwidth is especially useful to deal with spatial heterogeneity. Calibration of bandwidth is based on the method of Akaike information criterion (AIC) (Akaike 1974). The AIC method can find the best model among a group of the candidates that has the maximum likelihood:

$$\mathrm{AIC} = 2k - \ln(L) \quad , (3)$$

where *L* is the likelihood function of the model fitness, *k* is the number of parameters. Therefore, the best bandwidth parameter is the one with the minimum AIC value. The statistical summary report of the GWR model is shown in Table 6. Figure 4 maps the residuals of the GWR model. The mean AIC is 346.12 in the GWR model, achieving a slightly better goodness-of-fit. Similarly, the adjusted r-squared slightly increased from 0.31 (OLS) to 0.333 (GWR), which indicates that there is a marginal gain in explanatory power by the GWR model with the use of localized data samples. Figure 5 displays maps of the local r-squared. The figure shows that local r-squared derived by the GWR model decrease gradually from northwest to southeast in the study area.

Table 6: Geographically weighted regression statistical summary report

Variables	**Min**	**25 % Quartile**	**50 % Quartile**	**75 % Quartile**	**Max**
Intercept	−4.430	−2.933	−2.168	−1.796	−1.494
Intensity	−0.081	−0.019	0.049	0.171	0.391

Variables	Min	25 % Quartile	50 % Quartile	75 % Quartile	Max
Loss per capita	4.33E−06	4.71E−06	5.01E−06	5.4E−06	5.9E−06
Factor 1	−0.446	−0.422	−0.400	−0.366	−0.310
Factor 2	−0.048	0.039	0.126	0.225	0.345
Factor 3	−0.889	−0.845	−0.810	−0.781	−0.735
Factor 4	−0.057	−0.010	0.029	0.081	0.146
Akaike's information criterion	346.12				
Adjusted r-squared	0.333				

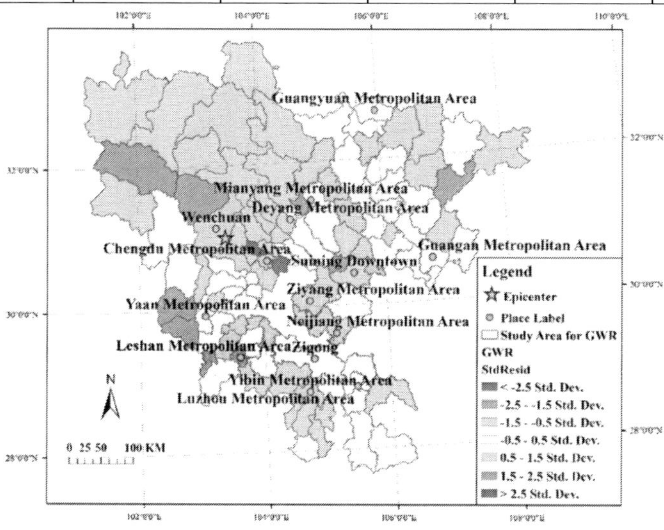

Fig. 4: Residuals of the geographically weighted regression model

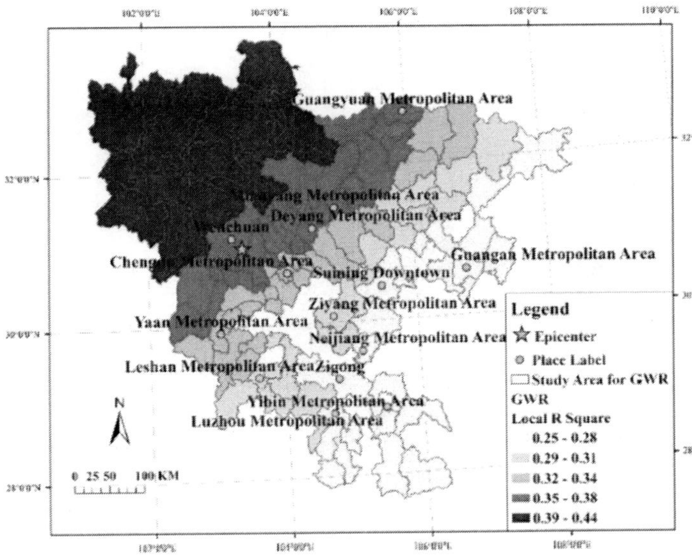

Fig. 5: Coefficients of determination (R-squared) from geographically weighted regression

Figure 6 shows the spatial patterns of the coefficients for intercept, intensity of earthquake, economic loss per capita, Factors 1–4. Local *p*-values were used to determine whether the coefficients were significant (shaded area) or not (non-shaded area). The intercepts are higher in the south than the north of the area. This distribution indicates the generally higher recovery capability of the counties in the southern part of Sichuan Province (Fig. 6a). Although the OLS model did not find any significant association between the earthquake intensity and resilience, the GWR model found the coefficients are significant for four counties located in the Guangyuan Metropolitan Area in northeast Sichuan (Fig. 6b). The coefficients of loss per capita are significant and positive for all the counties in the study area, but there is a downward spatial trend from the west towards the east (Fig. 6c). The coefficients of Factor 1 are the strongest near the metropolitan areas of Chengdu, Deyang, and Mianyang (Fig. 6d). The coefficients of Factor 2 are positive throughout the study area, but only are significant in a few southern areas such as Luzhou, Yibin, Zigong, and Leshan. The result is consistent with the global OLS regression. Similar to the OLS model, Factor 3 in the GWR model generally has negative coefficients. Nevertheless, the GWR model is able to describe the spatial distribution of the coefficients of Factor 3. As shown in Fig. 6f, the highest coefficients of Factor 3 are found in the southwestern part of the study area, especially in Yaan and Leshan areas, followed by Aba Prefecture. Also consistent with the OLS model, there are no significant coefficients of Factor 4 at the $p < 0.10$ level in the study area. Therefore, we do not include Factor 4 in our further discussion.

(a)

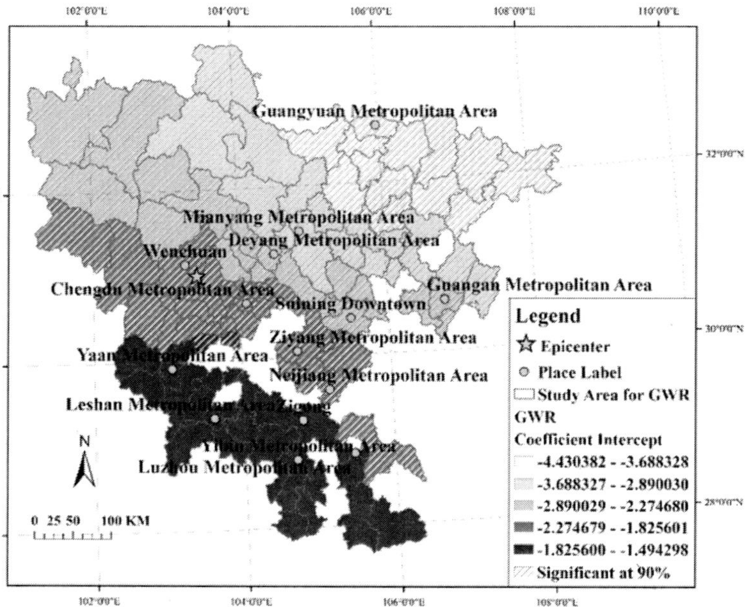

Intercept

(b)

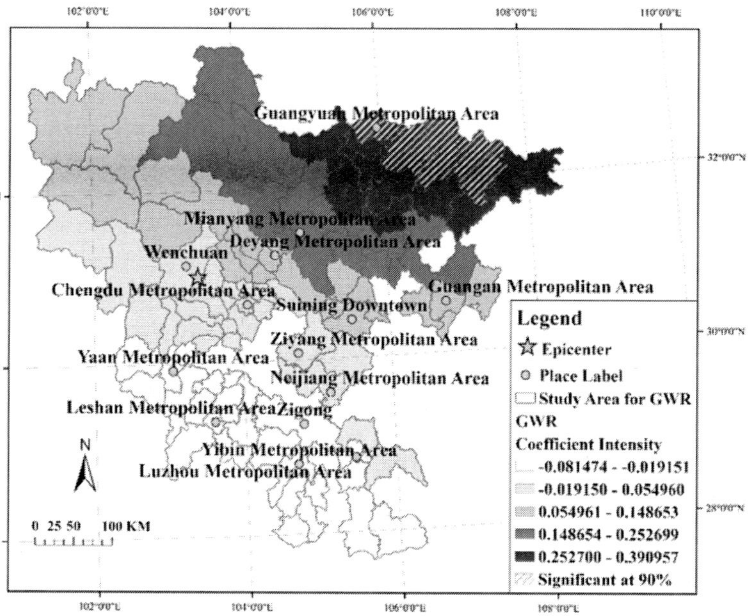

Intensity

(c)

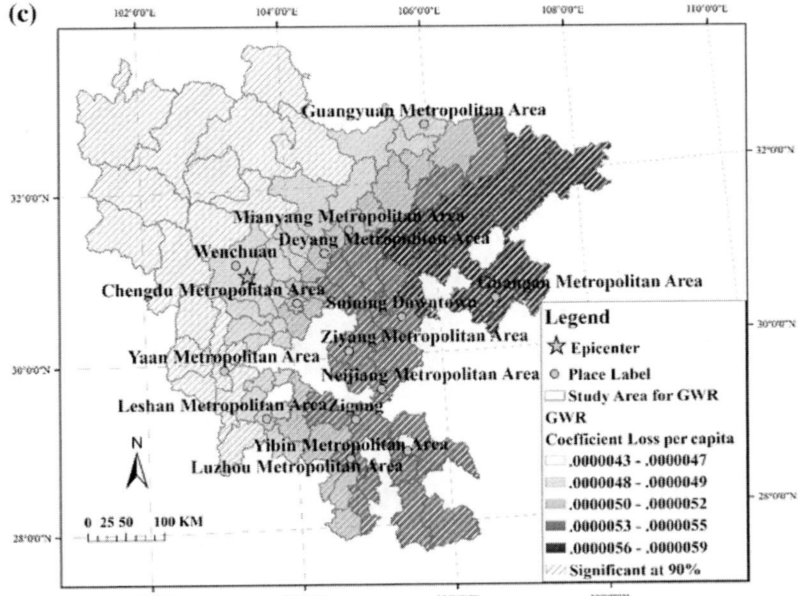

Loss per capita

(d)

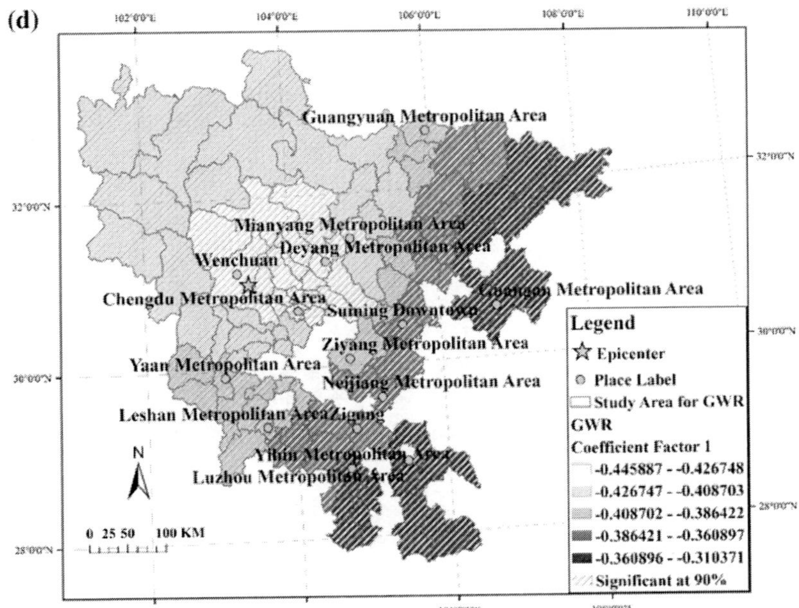

Factor 1

(e)

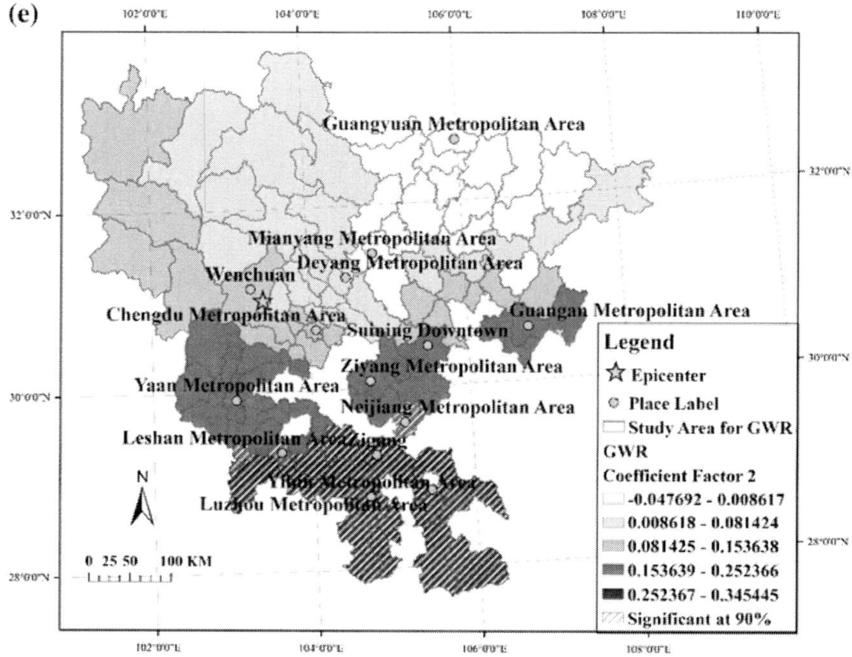

Factor 2

(f)

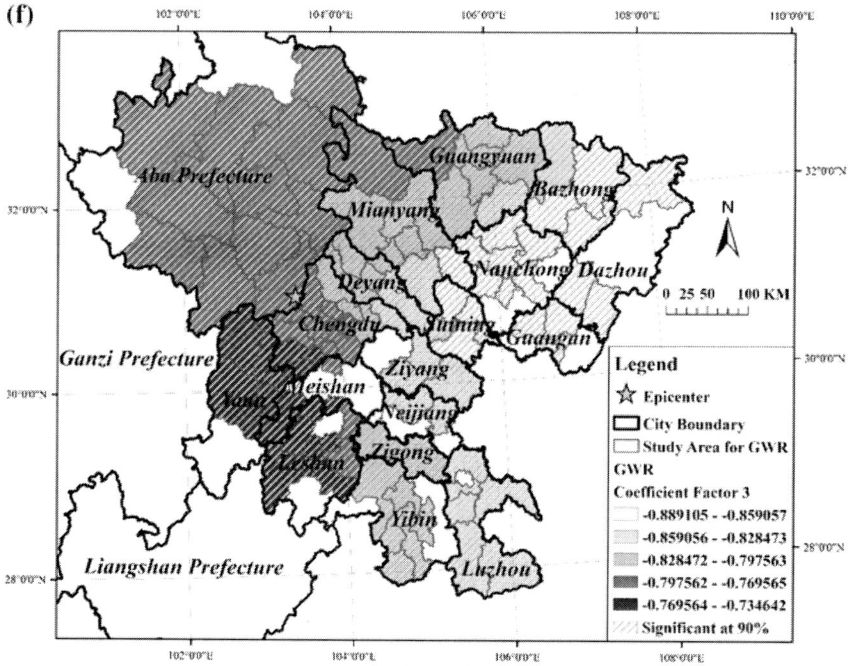

Factor 3

(g)

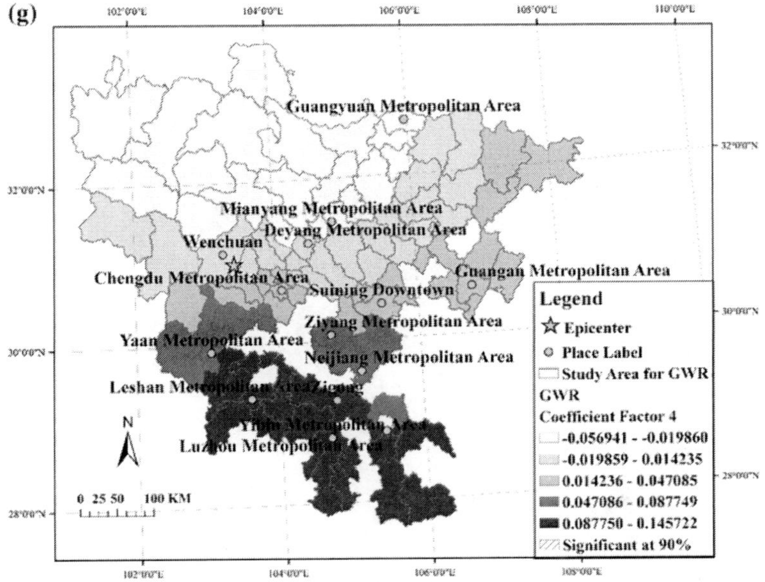

Factor 4

Fig. 6: GWR coefficients: **a** intercept, **b** earthquake intensity, **c** economic losses per capita, **d** Factor 1, **e** Factor 2, **f** Factor 3, and **g** Factor 4

3.5 Spatial Clusters of the GWR Coefficients

The coefficient maps produced by the GWR model show certain spatial continuities (Fig. 6), because the coefficients were optimized by local samples that are naturally spatially autocorrelated. Overlaying the coefficient maps for a cluster analysis would help identify boundaries of the underlying regionalization determined naturally by social and economic processes.

K-means cluster analysis (Hartigan and Wong 1979) is an unsupervised classification method commonly used to identify the natural groups (classes) among the data samples. The input variables are the standardized outputs from the GWR model, including the significant variable coefficients, their associated *p*-values, and the local r-squared. The number of preset classes (k) is an arbitrarily selected parameter for the K-means algorithm, which needs a user's specification or through optimization. A common practice of optimization is to select the optimal k value by using the AIC method (Milligan and Cooper 1985). In our analysis, the $k = 4$ gave the minimum AIC. Therefore, the study area was partitioned to four different regions by the K-means algorithm. The regions are shown in Fig. 7 and their statistics are summarized in Table 7. The mean values of the coefficients in each of the four regions depict the strength of the association of each independent variable with community resilience. The stronger the association, the better the possibility that relevant disaster relief policies will enhance resilience. Therefore, we can suggest following localized policies to improve community resilience to seismic hazards:

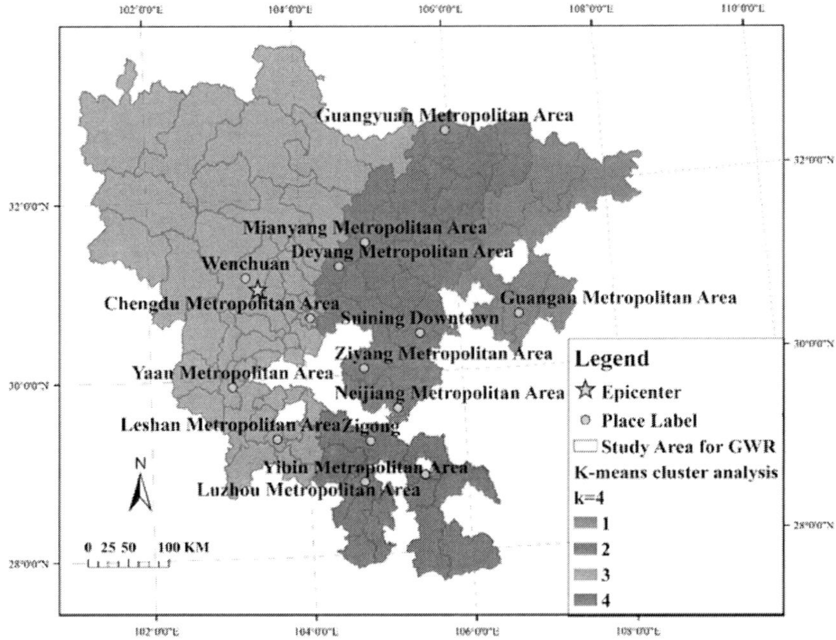

Fig. 7: Regions classified by K-means cluster analysis using local coefficients derived by GWR ($k = 4$)

1. Region 1 (the blue area in Fig. 7) mainly includes the counties in the northeastern part of the study area. In this region, intercept, loss per capita, and Factors 1 and 3 are significant. It has higher coefficients of intensity, loss per capita, and Factor 1, and lower coefficients of intercept and Factor 3 than other regions. Most counties are located relatively far from the epicenter of the earthquake. The GWR coefficients show that the negative association of the earthquake damage with the recovery rate was stronger than other regions. In order to achieve a better recovery rate in this area, it would be effective to allocate more recovery resources to those counties that suffered higher losses. The economic development level of this region is medium. The negative coefficients of Factor 3 (income level and ethnicity) lead to the conclusion that the minorities predominant in this area are less resilient to earthquake disasters than the general population. This is consistent with Cutter et al. (2003), that is, ethnic minority populations are most likely to be isolated by social or cultural barriers (Cutter et al. 2003, 2008), which would make them less resilient to disasters.

Table 7: Characteristics of clusters identified by the K-means cluster analysis

Regions	*n*	Mean of parameter estimates								
		Local R-squared	Coefficients				Local *p*-values			
			Intercept	Loss per capita	Factor 1	Factor 3	Intercept	Loss per capita	Factor 1	Factor 3
1	12	−1.13	−1.10	1.63	1.53	−1.64	−0.99	−1.12	1.69	0
2	13	−1.12	0.94	0.53	1.36	−0.38	1.38	−0.40	1.24	2.30
3	46	0.65	0.45	−0.91	−0.61	0.90	0.34	0.84	−0.58	−0.43
4	32	−0.06	−0.61	0.47	−0.25	−0.53	−0.68	−0.62	−0.31	−0.32
F-statistics*		64.6460	33.18	28.53	123.11	76.30	123.82	41.44	51.02	82.42
Sig.		0	0	0	0	0	0	0	0	0

* F statistics test: ANOVA test on the differences between the mean of local coefficients in the four clusters

2. Region 2 (the red area in Fig. 7) includes the counties in the southeastern segment of the study area. Here regression intercepts are not significant, an indication that this area has weak resilience in general. Factor 2 coefficients are positive and significant. Figure 6e reveals that Region 2 is unique in that it is the only area with significant Factor 2 coefficients. Factor 2 reflects economic dependency on secondary industries. The positive sign of Factor 2 suggests that the society supported by the secondary industries is more resilient to earthquake disasters than the general population. Other than Factor 2, this region has medium level coefficients of other factors among the regions.

Region 3 (the green area in Fig. 7) is located in the western half of the study area. In Region 3, four variables are significant: intercept, loss per capita, Factors 1 and 3. Among all regions, the resilience of Region 3 has the slightest influence from earthquake loss and the strongest influence from Factor 1—urbanization. In fact, this region suffered the most physical damage from the earthquake of the four regions. The socioeconomic situation of the local residents varied from well-to-do in Chengdu, Mianyang, and Deyang Metropolitan Areas to the underdeveloped surrounding rural areas occupied by non-Han ethnic minority populations. It is the dominant factor of reduced resilience. Language and cultural barriers as well as physical isolation caused by the topography made recovery difficult for those minority and low-income populations. More resources should be allocated to these regions when planning for disaster preparedness, relief, and recovery, than the general population.

Region 4 (the purple area in Fig. 7) constitutes the central transitional zone of the study area. In this region intercept, loss per capita, and Factors 1 and 3 are significant but the magnitudes are not very strong. Earthquake damage in this region has not had much impact on resilience. Overall, this region's ethnic minority population is a major negative impact factor on resilience. In other words, lower resilience is found in areas with more ethnic minority populations. Policymakers shall focus more on helping ethnic minority population improve their social and economic status.

4 DISCUSSION AND CONCLUSIONS

Although the result of the OLS regression analysis in this study confirms the relationship between the selected socioeconomic factors and community resilience to seismic hazard, the GWR model enables deeper exploration of geographical heterogeneity in the study area. General statistical methods have been criticized for their attempt to generalize human objects while neglecting the spatial structure of society. The use of a location-based model, GWR, compensates for the weakness of the pure statistical model that measures the correlation between a dependent variable and independent variables over the entire study area. For policy making, the spatial patterns of the coefficients can provide useful knowledge about local needs. In other words, the global regression answers generalized questions such as "Is factor *x* significant everywhere?", while the GWR model answers geographically specific questions such as "Is factor *x* significant here?" Location-based decisions can help allocate the limited resources available for disaster mitigation programs for their best use. Therefore, the use of geographical analysis based on local regression models such as GWR can effectively assist decision making in the disaster management sciences, especially in areas where the economic and social conditions of the society are highly variable by locations.

As a local regression model, GWR is able to account for the spatial continuity of sample data. We used GWR to explore the relationships between socioeconomic variables and resilience. However, several critiques of GWR exist. Most prominent are claims that GWR models lack a rigorous, formal statistical analysis (Congdon 2003; LeSage 2004; Wheeler and Calder 2007), and local collinearity (Wheeler and Tiefelsdorf 2005; Yu et al. 2009). We do not rate GWR models higher than traditional OLS regression models. Both model forms were helpful to our research in understanding the variables and problem to be explained. We used the OLS regression to confirm the selection of variables and GWR to explore the spatial pattern of the regression coefficients. Our tests showed that no local collinearity existed in our data, which was indicated by the small local condition numbers in the regression report.

Furthermore, the quantitative methods used in our study compare favorably with those used by other social science and humanity studies. We used the regionalization analysis to partition the study area into multiple sections characterized by relatively uniform coefficient values generated by the local regression model. This is similar to what has been used in regional geography research that studies the unique combination of characteristics of an area (Peet 1998). Despite the similarity in the form and the descriptive nature, our approach is fundamentally different from those of the traditional regional social science that were criticized for their lack of scientific evidence to define a "region." The regions defined by the traditional approach are subjective and scientifically unpredictable. In contrast, our approach is based on the quantitative methodology that underlies regression models—our regional divisions are empirically defined. Specifically, in this study we used the K-means clustering method to define regions. The regions created by our regionalization analysis are predictable and scientifically justifiable, and essentially it is the GIS that makes such an approach possible. Hence, one of the contributions of this article is to illustrate and promote the use of GIS-based spatial analysis and statistics in natural disaster studies.

A study of society's resilience to a major earthquake event was illustrated using regression models. In the study area, earthquakes are a tremendously destructive threat to the community and this geological hazard has occurred at fairly regular time intervals (Li et al. 2015b). The study area included over 105 counties in Sichuan, southwestern China, which were hit in 2008 by the Wenchuan Earthquake. A GWR model revealed the spatially varying strength of the area's socioeconomic factors in relation to the social resilience of local communities to seismic hazard. Based on spatial analysis in a GIS, we were able to infer location-sensitive decisions for allocation of disaster relief resources provided by the government. This invokes a new approach to spatial decision support in disaster management science. In this article, we also illustrated a general analytical framework for community resilience assessment. We anticipate that similar approaches will be applied to other disaster management studies. In future research, it would be beneficial to add a time dimension to our data for more comprehensive analyses, which enables longitudinal trend analyses of disaster events and the study of the coupled human-nature system and disaster-response system in a spatiotemporal framework.

ACKNOWLEDGMENTS

This material is partially based upon Project funded by the China Postdoctoral Science Foundation (Award No. 2016M592647). Any opinions, findings, and conclusions or recommendations expressed in this material are those of the authors and do not necessarily reflect the views of the funding agencies.

REFERENCES

1. Adger, W.N., T.P. Hughes, C. Folke, S.R. Carpenter, and J. Rockstrom. 2005. Social–ecological resilience to coastal disasters. *Science* 309(5737): 1036–1039.

2. Akaike, H. 1974. A new look at the statistical model identification. *IEEE Transactions on Automatic Control* AC-19(6): 716–723.

3. Bevington, J.S., A.A. Hill, R.A. Davidson, S.E. Chang, A. Vicini, B.J. Adams, and R.T. Eguchi. 2011. Measuring, monitoring, and evaluating post-disaster recovery: A key element in understanding community resilience. *Proceedings of ASCE Structures Congress 2011*, American Society of Civil Engineers, 2033–2043.

4. Brown, D.D., and J.C. Kulig. 1996. The concepts of resiliency: Theoretical lessons from community research. *Health and Canadian Society* 4(1): 30–50.

5. Bruneau, M., S.E. Chang, R.T. Eguchi, G.C. Lee, T.D. O'Rourke, A.M. Reinhorn, M. Shinozuka, K. Tierney, W.A. Wallace, and D. von Winterfeldt. 2003. A framework to quantitatively assess and enhance the seismic resilience of communities. *Earthquake Spectra* 19(4): 733–752.

6. Brunsdon, C., A.S. Fotheringham, and M.E. Charlton. 1998. Geographically weighted regression-modelling spatial non-

stationarity. *Journal of the Royal Statistical Society: Series D (The Statistician)* 47(3): 431–443.

7. Chang, S.E. 2010. Urban disaster recovery: A measurement framework and its application to the 1995 Kobe Earthquake. *Disasters* 34(2): 303–327.

8. Chatfield, C., and A.J. Collins. 1980. *Introduction to multivariate analysis*. London: Chapman and Hall.

9. Cloke, P., C. Philo, and D Sadler. 1991. *Approaching Human Geography*. London: Chapman.

10. Congdon, P.D. 2003. Modelling spatially varying impacts of socioeconomic predictors on mortality outcomes. *Journal of Geographical Systems* 5(2): 161–184.

11. Cutter, S.L., B.J. Boruff, and W.L. Shirley. 2003. Social vulnerability to environmental hazards. *Social Science Quarterly* 84(2): 242–261.

12. Cutter, S.L., L. Barnes, M. Berry, C. Burton, E. Evans, E. Tate, and J. Webb. 2008. A place-based model for understanding community resilience to natural disasters. *Global Environmental Change* 18(4): 598–606.

13. Cutter, S.L., C.G. Burton, and C.T. Emrich. 2010. Disaster resilience indicators for benchmarking baseline conditions. *Journal of Homeland Security and Emergency Management* 7(1). Article no. 51.

14. Finch, C., C.T. Emrich, and S.L. Cutter. 2010. Disaster disparities and differential recovery in New Orleans. *Population and Environment* 31(4): 179–202.

15. Fotheringham, A.S., P.J. Densham, and A. Curtis. 1995. The zone definition problem in location-allocation modeling. *Geographical Analysis* 27(1): 60–77.

16. Fotheringham, A.S., M.E. Charlton, and C. Brunsdon. 1998. Geographically weighted regression: A natural evolution of the expansion method for spatial data analysis. *Environment and Planning* A 30(11): 1905–1927.

17. Fotheringham, A., C. Bunsden, and M. Charlton. 2000. *Quantitative geography*. London: Sage.

18. Fotheringham, A.S., C. Brunsdon, and M.E. Charlton. 2003. *Geographically weighted regression*: *The analysis of spatially varying relationships*. London: Wiley.

19. Ghosh, D., and S.M. Manson. 2008. Robust principal component analysis and geographically weighted regression: Urbanization in the

twin cities metropolitan area of Minnesota. *Journal of the Urban and Regional Information Systems Association* 20(1): 15–25.

20. Guo, Y. 2012. Urban resilience in post-disaster reconstruction: Towards a resilient development in Sichuan, China. *International Journal of Disaster Risk Science* 3(1): 45–55.

21. Hair, J.F. Jr., R.E. Anderson, R.L. Tatham, and W.C. Black. 1998. *Multivariate data analysis*, 5th edn. Upper Saddle River, New Jersey: Prentice Hall.

22. Harman, H.H. 1976. *Modern factor analysis*. Chicago: University of Chicago Press.

23. Hartigan, J.A., and M.A. Wong. 1979. A K-means clustering algorithm. *Journal of the Royal Statistical Society, Series C (Applied Statistics)* 28(1): 100–108.

24. Hastie, T.J., and R.J. Tibshirani. 1990. *Generalized additive models*. New York: Chapman and Hall.

25. Holling, C.S. 1973. Resilience and stability of ecological systems. *Annual Review of Ecology and Systematics* 4: 1–23.

26. Holling, C.S. 1996. Engineering resilience versus ecological resilience. In *Engineering within Ecological Constraints*, ed. P.C. Schulze, 31–44. Washington, DC: National Academy Press.

27. Jolliffe, I.T. 2010. *Principal component analysis*, 2nd edn. New York: Springer.

28. Lam, N.S., M. Reams, K. Li, C. Li, and L.P. Mata. 2015. Measuring community resilience to coastal hazards along the Northern Gulf of Mexico. *Natural Hazards Review* 17(1). 04015013.

29. LeSage, J.P. 2004. A family of geographically weighted regression models. In *Advances in spatial econometrics: Methodology, tools and applications*, ed. L. Anselin, R. Florax, and S.J. Rey, 241–264. Berlin: Springer.

30. Li, W., C.A. Airriess, A.C.-C. Chen, K.J. Leong, and V. Keith. 2010. Katrina and migration: Evacuation and return by African Americans and Vietnamese Americans in an Eastern New Orleans suburb. *The Professional Geographer* 62(1): 103–118.

31. Li, X., N. Lam, Y. Qiang, K. Li, L. Yin, S. Liu, and W. Zheng. 2015a. Measuring county resilience after the 2008 Wenchuan Earthquake. *Natural Hazards and Earth System Sciences Discussions* 3: 81–122.

32. Li, X., W. Zheng, D. Wang, L. Yin, and Y. Wang. 2015b. Predicting seismicity trend in southwest of China based on wavelet analysis. *International Journal of Wavelets, Multiresolution and Information Processing* 13(2). Article no. 1550011: 1–9.

33. Milligan, G.W., and M.C. Cooper. 1985. An examination of procedures for determining the number of clusters in a data set. *Psychometrika* 50(2): 159–179.

34. National Bureau of Statistics of the People's Republic of China. 2001. *Tabulation of the 2000 population census of the People's Republic of China*. Beijing: China Statistics Press.

35. Nelson, E., G. Mendoza, J. Regetz, S. Polasky, H. Tallis, D. Cameron, K.M. Chan, G.C. Daily, et al. 2009. Modeling multiple ecosystem services, biodiversity conservation, commodity production, and tradeoffs at landscape scales. *Frontiers in Ecology and the Environment* 7: 4–11.

36. Norris, F.H., S.P. Stevens, B. Pfefferbaum, K.F. Wyche, and R.L. Pfefferbaum. 2008. Community resilience as a metaphor, theory, set of capacities, and strategy for disaster readiness. *American Journal of Community Psychology* 41(1–2): 127–150.

37. NRC (National Research Council). 2012. *Disaster resilience: A national imperative.* Washington, DC: The National Academies Press.

38. Panik, M.J. 2009. *Regression modeling: Methods, theory, and computation with SAS*. New York: Taylor and Francis Group.

39. Paton, D., and D.M. Johnston. 2001. Disasters and communities: Vulnerability, resilience and preparedness. *Disaster Prevention and Management: An International Journal* 10(4): 270–277.

40. Paton, D., and D.M. Johnston. 2006. *Disaster resilience: An integrated approach.* Springfield: Charles C. Thomas Publisher.

41. Paton, D., L. Smith, and J. Violanti. 2000. Disaster response: Risk, vulnerability and resilience. *Disaster Prevention and Management* 9(3): 173–180.

42. Peet, R. 1998. *Modern geographical thought*. Oxford: Blackwell.

43. Pelling, M. 2003. *The vulnerability of cities: Natural disasters and social resilience*. London and New York: Earthscan.

44. Shan, W. 2010. Comparing ethnic minorities and Han Chinese in China: Life satisfaction, economic well being and political attitudes. *East Asian Policy* 2: 13–22.

45. Sichuan Provincial Bureau of Statistics. 2003. *Sichuan statistical yearbook*. Beijing: China Statistics Press.

46. Sichuan Provincial Bureau of Statistics. 2006. *Sichuan statistical yearbook*. Beijing: China Statistics Press.

47. Sichuan Provincial Bureau of Statistics. 2008. *Sichuan statistical yearbook*. Beijing: China Statistics Press.

48. Sichuan Provincial Bureau of Statistics. 2009. *Sichuan statistical yearbook*. Beijing: China Statistics Press.

49. Sichuan Provincial Bureau of Statistics. 2011. *Sichuan statistical yearbook*. Beijing: China Statistics Press.

50. Smith, M. 1998. *Social science in question.* London: Sage.

51. The H. John Heinz III Center for Science, Economics, and the Environment. 2000. *The Hidden costs of coastal hazards: Implications for risk assessment and mitigation.* Washington, DC: Island Press.

52. Turner, B.L., R.E. Kasperson, P.A. Matson, J.J. McCarthy, R.W. Corell, L. Christensen, N. Eckley, J.X. Kasperson, A. Luers, M.L. Martello, C. Polsky, A. Pulsipher, and A. Schiller. 2003. A framework for vulnerability analysis in sustainability science. *Proceedings of the National Academy of Sciences* 100(14): 8074–8079.

53. UNISDR (United Nations International Strategy for Disaster Reduction). 2004. *Living with risk: A global review of disaster reduction initiatives.* Geneva: UNISDR.

54. Unwin, D.J. 1996. GIS, spatial analysis and spatial statistics. *Progress in Human Geography* 20(4): 540–551.

55. USGS (United States Geological Survey). 2008. *Magnitude 7.9*: *Eastern Sichuan, China*. U.S. Geological Survey Earthquake Hazards Program. http://earthquake.usgs.gov/earthquakes/eqinthenews/2008/us2008ryan/. Accessed 27 May 2016.

56. Wheeler, D.C., and C.A. Calder. 2007. An assessment of coefficient accuracy in linear regression models with spatially varying coefficients. *Journal of Geographical Systems* 9(2): 145–166.

57. Wheeler, D.C., and M. Tiefelsdorf. 2005. Multicollinearity and correlation among local regression coefficients in geographically weighted regression. *Journal of Geographical Systems* 7(2): 161–187.

58. Yu, D., N.A. Peterson, and R.J. Reid. 2009. Exploring the impact of non-normality on spatial non-stationarity in geographically weighted regression analyses: Tobacco outlet density in New Jersey. *GIScience and Remote Sensing* 46(3): 329–346.

59. Zheng, W., X. Li, N. Lam, X. Wang, S. Liu, X. Yu, Z. Sun, and J. Yao. 2013. Applications of integrated geophysical method in archaeological surveys of the ancient Shu ruins. *Journal of Archaeological Science* 40(1): 166–175.

60. Zheng, W., X. Li, J. Xie, L. Yin, and Y. Wang. 2015. Impact of human activities on haze in Beijing based on grey relational analysis. *Rendiconti Lincei* 26(2): 187–192.

61. Zheng, W., X. Li, L. Yin, and Y. Wang. 2016. Spatiotemporal heterogeneity of urban air pollution in China based on spatial analysis. *Rendiconti Lincei* 27(2): 351–356.

INDEX

P

R

S

U

W